AF360884

Advances in the Molecular Mechanisms of Abscisic Acid and Gibberellins Functions in Plants

Advances in the Molecular Mechanisms of Abscisic Acid and Gibberellins Functions in Plants

Editor

Víctor Quesada

MDPI • Basel • Beijing • Wuhan • Barcelona • Belgrade • Manchester • Tokyo • Cluj • Tianjin

Editor
Víctor Quesada
Bioengineering Institute
Universidad Miguel Hernández
de Elche
Elche
Spain

Editorial Office
MDPI
St. Alban-Anlage 66
4052 Basel, Switzerland

This is a reprint of articles from the Special Issue published online in the open access journal *IJMS* (ISSN 1422-0067) (available at: www.mdpi.com/journal/ijms/special_issues/AAAPlant).

For citation purposes, cite each article independently as indicated on the article page online and as indicated below:

LastName, A.A.; LastName, B.B.; LastName, C.C. Article Title. *Journal Name* **Year**, *Volume Number*, Page Range.

ISBN 978-3-0365-1540-3 (Hbk)
ISBN 978-3-0365-1539-7 (PDF)

Contents

About the Editor

Víctor Quesada

Dr. Quesada is an Associate Professor at the Universidad Miguel Hernández de Elche (UMH), Spain. He completed his undergraduate studies in Biology at the Universidad de Alicante, Spain, in 1993 and received his Ph.D. at the UMH working in the isolation and characterization of salt-tolerant mutants in the plant model system *Arabidopsis thaliana* (1999). He moved then to the John Innes Centre (JIC) in Norwich, UK, from 2000 to 2003, to work in flowering time regulation in Arabidopsis. After finishing his post-doc, he moved to the Bioengineering Institute of the UMH to work in leaf morphogenesis. He became an Assistant Professor (December 2003) and an Associate Professor in 2009 at the UMH. Since 2009, his independent research activity has been focused on the functional characterization of Arabidopsis nuclear genes involved in organellar gene expression to shed light on the understanding of their roles in plant growth, development and adaptation to abiotic stress.

International Journal of
Molecular Sciences

Editorial

Advances in the Molecular Mechanisms of Abscisic Acid and Gibberellins Functions in Plants

Víctor Quesada 

Instituto de Bioingeniería, Universidad Miguel Hernández, Campus de Elche, 03202 Elche, Spain; vquesada@umh.es; Tel.: +34-96-665-8812

Citation: Quesada, V. Advances in the Molecular Mechanisms of Abscisic Acid and Gibberellins Functions in Plants. *Int. J. Mol. Sci.* **2021**, *22*, 6080. https://doi.org/10.3390/ijms22116080

Received: 25 May 2021
Accepted: 31 May 2021
Published: 4 June 2021

Publisher's Note: MDPI stays neutral with regard to jurisdictional claims in published maps and institutional affiliations.

In this special issue entitled, *"Advances in the Molecular Mechanisms of Abscisic Acid and Gibberellins Functions in Plants"*, eight articles are collected, with five reviews and three original research papers, which broadly cover different topics on the abscisic acid (ABA) field and, to a lesser extent, on gibberellins (GAs) research. These works explore ABA involvement in processes like flowering, plant defense, abiotic stress response or maturation of non-climacteric fruits, with reports in the last case of interplay between ABA and GAs. New findings on the regulation of ABA or GAs activity are also reported in this issue. The experimental studies and reviews published in this special issue focus on the results obtained using principally the plant model system *Arabidopsis thaliana*. Notwithstanding, other works based on agronomically important plants, such as grapevine or citrus species, are also included, and reveal the crucial role of ABA and GAs across different plant species. I summarize here the main findings of these works, which represent outstanding advances in our knowledge on the molecular mechanisms through which ABA and GAs control fundamental processes in plants.

Phytohormones GAs and ABA antagonistically regulate both plant growth and several developmental processes, such as seed maturation, seed dormancy, germination, hypocotyl elongation, primary root growth and flowering time. ABA and GAs generally inhibit and promote respectively cell elongation and growth, and the mutual antagonism between these two phytohormones governs many developmental decisions in plants [1].

In addition to the growing body of evidence for ABA as a modulator of plant growth and development [2], ABA is primarily known for being a fundamental player in the response, tolerance and adaptation of plants to diverse abiotic stress conditions, among which low temperatures, heat, drought, salinity or flooding are highlighted [3]. Interestingly, different recent works suggest a function for GAs in controlling some biological processes in response to stress [4–6].

This special issue "Advances in the Molecular Mechanisms of Abscisic Acid and Gibberellins Functions in Plants" contains eight articles; most of these focus on ABA, five are review articles and three are original research papers published by field experts. These manuscripts will help to understand the fundamental roles of GAs and ABA in the regulation of plant growth, development, and in responses to abiotic or biotic stresses. These articles will also shed light on the molecular mechanisms of ABA and GAs action in plants. The reviews published in this issue focus on the interaction between ABA and GAs in regulating non climacteric fruit development and maturation [7], current knowledge on the role of ABA in mediating mechanisms whereby grapevine deals with abiotic stresses [8], the analysis of the role of ABA in flowering transition [9], the mechanism by which the type 2C Protein Phosphatases (PP2C) gene transcription modulates ABA signaling [10], and the role of the Mediator complex on the ABA signaling pathway and abiotic stress response [11]. The original research articles investigate the causes and consequences of the kaolin-induced modulation of ABA biosynthesis in grapevines when faced with a water deficit [12], the involvement of ABA in plant immune responses [13], and the relation between GAs and Arabidopsis receptor-like cytoplasmic kinase (RLCK) VI_A2 [14]. In this editorial, I summarize the main findings of these eight insightful works.

Fruit development and maturation results from an intricate interplay of molecular and physiological processes, regulated by endogenous (hormones) and external (environment) factors. The paper by Alferez et al. [7] exhaustively reviews the interplay between ABA and GAs in regulating the development and maturation of non-climacteric fruits at the molecular level, in economically important fruits like grape berries, strawberries and citruses. Alferez et al. [7] also relate the interaction of ABA and GAs with ethylene and sugar signaling in modulating non-climacteric fruit development and maturation. They depict a time-course model in which GA levels lower, ABA levels rise and ethylene production remains steady, whereas ethylene perception increases as non-climacteric fruit maturation progresses. Notwithstanding, Alferez et al. [7] also highlight that although current knowledge clearly points out the crosstalk between ABA and GAs as a major factor that controls fruit maturation, fine details of this regulation are still not well-understood and warrant further research. Finally, the authors examine the increasing body of evidence about ABA, GAs and the genus Citrus, which reveal that this woody genus can be considered an emerging plant model system for non-climacteric maturation studies.

Climate change poses a threat to important agricultural regions in the world, such as Mediterranean-climate areas, where socio-economically relevant crops (e.g., grapevine) can be seriously threatened. Consequently, improvements in viticulture techniques are needed, and better knowledge of grapevine physiology under stress conditions is required to achieve this. Besides its socio-economic importance, *Vitis vinifera* is also a model species in drought-response research. Marusig and Tombesi [8] provide an exhaustive overview of current knowledge on the role of ABA in mediating mechanisms whereby grapevines cope with abiotic stresses. In line with this, these authors especially focus on the mechanisms of ABA biosynthesis and translocation, the role of this phytohormone in regulating stomata closure and carbohydrates mobilization in response to drought stress, as well as ABA involvement in salt stress. The results of all these works clearly demonstrate that in *Vitis vinifera*, ABA is a key hormone involved in regulating the mechanisms for coping with major threats caused by climate change. Marusig and Tombesi [8] also highlight some main issues that deserve further research in this field, which are fundamentally the understanding of the role that ABA plays in drought stress in relation to water stress severity, duration and frequency, the interaction of ABA regulation and carbohydrates under water stress, and a profounder knowledge of the ABA function in the salt stress response.

Drought escape (DE), which is accelerated flowering that plants undergo in the vegetative phase in response to a water deficit, is considered an adaptive strategy for survival in dry climates. Studying DE in Arabidopsis has revealed that ABA plays a prominent role in controlling floral transition. The review by Martignago et al. [9] thoroughly explores current knowledge on how and in what spatial context ABA signals can affect the intricate floral network. These authors review ABA signaling and its multiple connections with the photoperiodic pathway, as the DE process is highly intertwined with photoperiodic genes, and integrating drought stimuli into the floral network is mediated largely by ABA. Martignago et al. [9] emphasize the important role of the *GIGANTEA* (*GI*) gene in integrating ABA signaling into flowering. *GI* is a key flowering gene required for photoperiod perception and clock function, and it also emerges as the key driver of DE. The current model suggests that the transcription factors (TFs) activated by ABA would recruit GI in different genomic positions to regulate gene networks involved in the drought stress response. These authors also analyze the putative role of FD and FD-like TFs in the modulation of ABA responses through interactions with FLOWERING LOCUS T (FT) and FT-like proteins at the shoot apex. Finally, Martignago et al. [9] discuss how the progress made in Arabidopsis about ABA-flowering molecular interactions can be transferred to crops. Along these lines, they highlight the utility of studying DE traits in natural populations, and the importance of rice as the DE pathway is essentially conserved in this species.

ABA plays a pivotal role in controlling plant stomata closure in response to osmotic stress, which prevents water loss. However even under stressful conditions, stomatal apertures occur to uptake CO_2 for photosynthesis. Consequently, ABA levels must be very fine-tuned regulated to maintain plant homeostasis, which is achieved by controlling this hormone's biosynthesis, catabolism and signaling pathway. Jung et al. [10] comprehensively review how plants modulate the ABA signaling pathway by focusing on the transcriptional regulation of *PP2C* gene expression by ABA. They report that plant *PP2Cs* are a fundamental switch at the core of the ABA signaling network. Jung et al. [10] summarize the results hitherto published and demonstrate that ABA induces both repressors and activators of *PP2C* gene transcription to modulate ABA responses. These regulators can also affect the chromatin state by thereby regulating the transcription of ABA-responsive genes. Remarkably, this stress-induced chromatin remodeling state can be memorized, and even inherited, by the next generation of plants.

The Mediator is a conserved eukaryotic multiprotein complex that modulates the association between transcription factors and RNA-polymerase II to accurately regulate gene transcription. The functions of the Mediator complex in plant development processes and the biotic stress response have been extensively studied. However, its roles on the ABA signaling pathway and abiotic stress responses are still poorly known. Chong et al. [11] exhaustively summarize current knowledge on the regulatory roles of the Mediator complex on the ABA signaling pathway and in plant responses to three abiotic stresses: cold, high salinity and drought. These authors show that the Mediator complex is critical for plants to respond to abiotic stresses, and particularly focus on the participation of Mediator subunits MED16, MED18, MED25 and CDK8 in response to ABA and environmental cues. Chong et al. [11] report that Mediator subunits display multifunctional roles in salt and drought stress, and they discuss further potential research approaches needed to ascertain the role of the Mediator complex in regulating ABA and abiotic stress responses.

Kaolin is a natural clay used in some crops to alleviate the negative impact of extreme temperatures and/or drought on leaves and fruits, due to its light reflective properties. In line with this, a reduction in leaf ABA content associated with better leaf stomatal conductance induced by kaolin has been reported in grapevines. The research manuscript by Frioni et al. [12] explores the causes and consequences of not only the kaolin-induced modulation of ABA biosynthesis under progressive water shortage, but also the dynamic interactions between kaolin and ABA precursors violaxanthin (Vx), antheraxanthin (Ax), zeaxanthin (Zx) (representing the molecules involved in the xanthophyll (VAZ) cycle) and neoxanthin (Nx). Frioni et al. [12] report that kaolin, under water deficit, preserves leaf transpiration and reduces the accumulation of ABA in grapevine leaves by avoiding the deviation of the VAZ epoxidation/de-epoxidation cycle in the biosynthesis of Nx. Their findings contribute to explaining the mechanisms involved in the kaolin-induced protection of canopy functionality.

In addition to the aforementioned role of ABA in both controlling plant responses to abiotic stress and regulating plant developmental processes and growth, this hormone is also involved in responses to biotic stresses caused by a wide range of plant pathogens. ABA's role in plant pathogen response is complex, given its interplay with key players in defense (e.g., salicylic acid (SA), jasmonic acid (JA) and ethylene (ET)), and also because ABA outcomes depend on the biology of the infective pathogen. Furthermore, the specific receptor of ABA, which activates the positive or negative ABA responses during immune responses, has to date remained unknown. The comprehensive study by García-Andrade et al. [13] unveils a non-redundant role in plant immunity for one of the 14 multigene ABA receptor family members, namely ABA receptor PYRABACTIN RESISTANCE 1 (PYR1). This research manuscript reveals that this receptor is crucial for modulating the cross-talk between the SA and ET signaling pathways in plant defense. The results of García-Andrade et al. [13] also demonstrate that ABA-activated SNF1-related protein kinases (SnRKs) subfamily 2 (SnRK2s) are fundamental components for plant resistance to pathogens.

Valkai et al. [14] investigate the biological function of the Arabidopsis *RLCK VI_A2* gene in the regulation of plant growth and skotomorphogenesis. RLCK VI_A2 is a member of the plant-specific receptor-like cytoplasmic kinases (RLCKs) that form a large and barely characterized family. The activity of the RLCK VI_A class of dicots is regulated by Rho-of-plants (ROP) GTPases. Valkai et al. [14] show that loss of the *RLCK VI_A2* function leads to reduced cell expansion and seedling growth. These mutant phenotypes can be rescued by the exogenous application of GAs. However, differences in neither GA content nor GA sensitivity of the *RLCK VI_A2* defective mutant have been found compared to the wild type. An RNA-seq analysis indicated that the RLCK VI_A2 kinase and GAs can act in parallel to regulate cell expansion and plant growth. Interestingly, the transcriptomic analysis also revealed a role for RLCK VI_A2 kinase in cellular transport and cell wall organization.

Altogether, the contributions published in this special issue are excellent examples of the recent advances made in the molecular mechanisms of ABA and GA functions in plants. I wish to thank all the authors for their contributions and the reviewers for their critical assessments of these articles. I also thank the assistant editor Ms. Reyna Li for providing me with the opportunity to serve as the Guest Editor of this special issue.

Funding: This research received no external funding.

Conflicts of Interest: The author declares no conflict of interest.

References

1. Shu, K.; Zhou, W.; Chen, F.; Luo, X.; Yang, W. Abscisic Acid and Gibberellins Antagonistically Mediate Plant Development and Abiotic Stress Responses. *Front Plant Sci.* **2018**, *9*, 416. [CrossRef] [PubMed]
2. Zhu, J.K. Salt and drought stress signal transduction in plants. *Annu. Rev. Plant Biol.* **2002**, *53*, 247–273. [CrossRef] [PubMed]
3. Yoshida, T.; Christmann, A.; Yamaguchi-Shinozaki, K.; Grill, E.; Fernie, A.R. Revisiting the Basal Role of ABA—Roles Outside of Stress. *Trends Plant Sci.* **2019**, *24*, 625–635. [CrossRef] [PubMed]
4. Hamayun, M.; Hussain, A.; Khan, S.A.; Kim, H.Y.; Khan, A.L.; Waqas, M.; Irshad, M.; Iqbal, A.; Rehman, G.; Jan, S.; et al. Gibberellins producing endophytic fungus Porostereum spadiceum AGH786 rescues growth of salt affected soybean. *Front. Microbiol.* **2017**, *8*, 686. [CrossRef] [PubMed]
5. Urano, K.; Maruyama, K.; Jikumaru, Y.; Kamiya, Y.; Yamaguchi-Shinozaki, K.; Shinozaki, K. Analysis of plant hormone profiles in response to moderate dehydration stress. *Plant J.* **2017**, *90*, 17–36. [CrossRef] [PubMed]
6. Wang, B.; Wei, H.; Xue, Z.; Zhang, W.H. Gibberellins regulate iron deficiency-response by influencing iron transport and translocation in rice seedlings (*Oryza sativa*). *Ann. Bot.* **2017**, *119*, 945–956. [CrossRef] [PubMed]
7. Alferez, F.; Uilian de Carvalho, D.; Boakye, D. Interplay between Abscisic Acid and Gibberellins, as Related to Ethylene and Sugars, in Regulating Maturation of Non-Climacteric Fruit. *Int. J. Mol. Sci.* **2021**, *22*, 669. [CrossRef] [PubMed]
8. Marusig, D.; Tombesi, S. Abscisic Acid Mediates Drought and Salt Stress Responses in *Vitis vinifera*—A Review. *Int. J. Mol. Sci.* **2020**, *21*, 8648. [CrossRef] [PubMed]
9. Martignago, D.; Siemiatkowska, B.; Lombardi, A.; Conti, L. Abscisic Acid and Flowering Regulation: Many Targets, Different Places. *Int. J. Mol. Sci.* **2020**, *21*, 9700. [CrossRef] [PubMed]
10. Jung, C.; Nguyen, N.H.; Cheong, J.J. Transcriptional Regulation of Protein Phosphatase 2C Genes to Modulate Abscisic Acid Signaling. *Int. J. Mol. Sci.* **2020**, *21*, 9517. [CrossRef] [PubMed]
11. Chong, L.; Guo, P.; Zhu, Y. Mediator Complex: A Pivotal Regulator of ABA Signaling Pathway and Abiotic Stress Response in Plants. *Int. J. Mol. Sci.* **2020**, *21*, 7755. [CrossRef] [PubMed]
12. Frioni, T.; Tombesi, S.; Sabbatini, P.; Squeri, C.; Rodas, N.L.; Palliotti, A.; Poni, S. Kaolin Reduces ABA Biosynthesis through the Inhibition of Neoxanthin Synthesis in Grapevines under Water Deficit. *Int. J. Mol. Sci.* **2020**, *21*, 4950. [CrossRef] [PubMed]
13. García-Andrade, J.; González, B.; González-Guzman, M.; Rodriguez, P.L.; Vera, P. The Role of ABA in Plant Immunity is Mediated through the PYR1 Receptor. *Int. J. Mol. Sci.* **2020**, *21*, 5852. [CrossRef] [PubMed]
14. Valkai, I.; Kénesi, E.; Domonkos, I.; Ayaydin, F.; Tarkowská, D.; Strnad, M.; Faragó, A.; Bodai, L.; Fehér, A. The Arabidopsis RLCK VI_A2 Kinase Controls Seedling and Plant Growth in Parallel with Gibberellin. *Int. J. Mol. Sci.* **2020**, *21*, 7266. [CrossRef] [PubMed]

International Journal of
Molecular Sciences

Review

Interplay between Abscisic Acid and Gibberellins, as Related to Ethylene and Sugars, in Regulating Maturation of Non-Climacteric Fruit

Fernando Alferez [1,*], Deived Uilian de Carvalho [1,2] and Daniel Boakye [1]

1 Southwest Florida Research and Education Center, Department of Horticulture, University of Florida–Institute of Food and Agricultural Sciences (UF–IFAS), Immokalee, FL 34142, USA; deived10@gmail.com (D.U.d.C.); dadu.boakye@ufl.edu (D.B.)
2 AC Jardim Bandeirante, Centro de Ciências Agrárias, Universidade Estadual de Londrina, Jardim Portal de Versalhes 1 86057970, Londrina/PR 10011, Brazil
* Correspondence: alferez@ufl.edu; Tel.: +239-658-3426; Fax: +239-658-3403

Abstract: In this review, we address the interaction between abscisic acid (ABA) and gibberellins (GAs) in regulating non-climacteric fruit development and maturation at the molecular level. We review the interplay of both plant growth regulators in regulating these processes in several fruit of economic importance such as grape berries, strawberry, and citrus, and show how understanding this interaction has resulted in useful agronomic management techniques. We then relate the interplay of both hormones with ethylene and other endogenous factors, such as sugar signaling. We finally review the growing knowledge related to abscisic acid, gibberellins, and the genus Citrus. We illustrate why this woody genus can be considered as an emerging model plant for understanding hormonal circuits in regulating different processes, as most of the finest work on this matter in recent years has been performed by using different Citrus species.

Keywords: abscisic acid; citrus; ethylene; fruit maturation; gibberellins; hormonal interplay; sugars

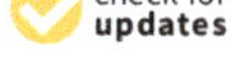

Citation: Alferez, F.; de Carvalho, D.U.; Boakye, D. Interplay between Abscisic Acid and Gibberellins, as Related to Ethylene and Sugars, in Regulating Maturation of Non-Climacteric Fruit. *Int. J. Mol. Sci.* **2021**, *22*, 669. https://doi.org/10.3390/ijms22020669

Received: 17 December 2020
Accepted: 10 January 2021
Published: 12 January 2021

Publisher's Note: MDPI stays neutral with regard to jurisdictional claims in published maps and institutional affiliations.

1. Introduction

The development and maturation of fruit is the result of a complex interplay of molecular, biochemical, and physiological processes, modulated by internal factors such as hormones and external factors such as the environment. In general, the transition from fruit growth to maturation involves changes in the sugar metabolism, and the softening and coloration of different fruit tissues. The development of fleshy fruit is divided into three distinct stages: the first stage (1) is recognized by slow growth as cell division takes place; the second stage (2) is marked by a rapid fruit growth due to cell expansion, and major increases in size and weight are observed; fruit ripening is initiated at the third stage (3), when fruit growth ceases and there is an increase in the biochemical reactions that result in fruit maturation involving fruit color-change, acid degradation, sugar accumulation, and other processes that combined result in final organoleptic attributes. Classically, fleshy fruits are classified into two physiological categories based on the respiration pattern and ethylene biosynthesis occurring at stage 3: climacteric and non-climacteric [1]. Climacteric fruits, such as tomato, apple, apricot, atemoya, banana, blueberry, guava, mango, papaya, and peach have an increase in the respiration rate and ethylene production at stage 3 when fruit ripening process enable fruit harvest prior to complete fruit maturation. On the other hand, non-climacteric fruits including strawberry, citrus, grape, cherry, plum, litchi, and others display a progressive reduction in the respiration rate during maturation while the ethylene production remains at basal level. The hormonal regulation of fruit ripening in climacteric fruit has been widely addressed, and in-depth studies, taking into account molecular aspects of hormone crosstalk and interaction with environment are abundant, greatly thanks to research on tomato, a very well characterized model plant due to a wide collection of mutants available and a faster growth cycle [2–4]. Hormonal interaction in regulating maturation of non-climacteric fruit has been also studied in several fruits including strawberry, grape berries and citrus among others. Whereas observational studies are relatively abundant, in-depth molecular and mechanistic studies have been performed only in a handful of fruit including strawberry, grape berries and citrus, mostly because of their economic importance. However, the focus of these kind of studies have been primarily on ethylene and its interaction with other hormones. In this review, we want to focus on interaction between abscisic acid (ABA) and gibberellins (GAs) in regulating non-climacteric fruit development and maturation at the molecular level. We then relate the interplay of both hormones with ethylene and other endogenous factors, such as sugar signaling. We finally review the growing knowledge related to abscisic acid, gibberellins, and the genus Citrus. We highlight how Citrus can be considered as an emerging model plant for understanding hormonal circuits in regulating different physiological processes, including fruit maturation and responses to stress, as most of the finest work on this matter in the last years has been performed by using different Citrus species.

2. Introduction to Abscisic Acid in Fruit

2.1. Abscisic Acid Biosynthesis and Accumulation

Accumulation of ABA in both climacteric and non-climacteric fruit during maturation is known for decades [5–9]. ABA is a product of the carotenoid pathway [5,6]. This plant hormone is derived from C40-*cis*-epoxycarotenoids, which are cleaved by the 9-*cis*-epoxycarotenoid dioxygenase (NCED) to produce xanthoxin, the direct C15 precursor of ABA [7–9]. Several studies in non-climacteric fruit have shown the role of this hormone in regulating the process of maturation. In cherries, endogenous ABA levels are the result of a balance between biosynthesis mediated by *PacNCEDs*, and catabolism mediated by *PacCYP707As* (encoding a 8′-hydroxylase, a key enzyme in the oxidative catabolism of ABA), and transcriptional regulation of these genes influence maturation [10,11]. In mangosteens, ABA accumulation in fruit peel and aril precedes fruit coloration and decrease in peel firmness, suggesting the involvement of this hormone in triggering maturation [12]. In Citrus, increase in ABA concentration in the fruit occurs during maturation in different species [13–15] and irrespective of the fruit tissue [16]. Lowering levels of ABA are accompanied by a delay in color change in lemons, and retarding senescence has been related also to lower levels of this hormone [17,18]. In addition, it has been noted a relation between ABA increase and transition from chloroplast to chromoplast in mandarin [19] and sweet cherry [10]. In Citrus, ABA increases in response to ethylene [19,20], and accumulates during fruit development, maturation and senescence [13,17].

2.2. Abscisic Acid Function During Fruit Maturation

Exogenous treatments with ABA may also have an effect in different maturation parameters, although there are some disparities in the response depending on the fruit. For instance, in field-grown grape berries, exogenous ABA increases maturation-related pigments such as anthocyanin and flavonol [21–23], advances the process of color change (veraison), and downregulates expression of genes associated with photosynthesis [24]. Interestingly, ABA may exert different actions on maturation and on ethylene biosynthesis depending on the stage of fruit development in grape [25].

The involvement of ethylene in non-climacteric maturation is not the focus of this review as there are many in-depth studies of these interactions [26] and will be addressed in Section 4, Integrating Signals to Regulate Maturation: GA, ABA, Sugars, and Ethylene Interaction. However, in the context of this article, it is worth to mention that combined application of ABA and the ethylene releasing compound ethrel to *Litchi chinensis* three weeks before harvest was more effective in enhancing both chlorophyll degradation and anthocyanin biosynthesis than the application of ABA alone, showing a possible synergistic effect of ABA and ethylene in promoting anthocyanin synthesis, chlorophyll degradation and ultimately peel coloration. Interestingly, in this study, exogenous ABA also induced sugar accumulation [27]. Exogenous application of ABA before color break also improved color in mandarin fruit (*Citrus reticulata* Blanco cv. Ponkan) [28] and in M7 sweet orange [29]; however, in other citrus fruit exogenous ABA did not promote color development [30], whereas in juice sacs cultured in vitro, ABA induced its own biosynthesis at the transcriptional level, and this feedback regulation of ABA led to a decrease in carotenoid content [31]. The nature of ABA synthesis, being a final product of the carotenoid biosynthetic pathway, makes particularly difficult to unravel and ultimately understand its role in coloration when carotenoids are the main pigments involved. This elusiveness can be largely avoided in fruit such as grapes, accumulating other classes of pigments, such as anthocyanins, as we discussed above.

3. GAs, ABA and Their Interplay during Fruit Development and Maturation

3.1. Integration of ABA and GAs Biosynthesis

ABA and GAs share their biosynthetic pathway with other plant growth regulators, including cytokinin (CK) and diverse sterols (Figure 1). In model plants, several GAs and ABA mutants have been identified and characterized, allowing the elucidation of their biosynthetic pathway and function. For example, *flacca* and *sitiens* mutants of tomato are defective in the last steps of carotenoids biosynthesis, thus impairing ABA synthesis with downstream effects [9]. In corn, studies on several viviparous mutants helped to elucidate the biochemistry of carotenoid biosynthesis; these mutants show blockages at different steps of the pathway, resulting in accumulation of precursors and reduction of ABA content [32,33]. Certainly, in woody plants such as Citrus the use of mutants altered in hormonal biosynthesis is much less feasible as artificially induced mutants are difficult to generate due to cost and lack of facile methods.

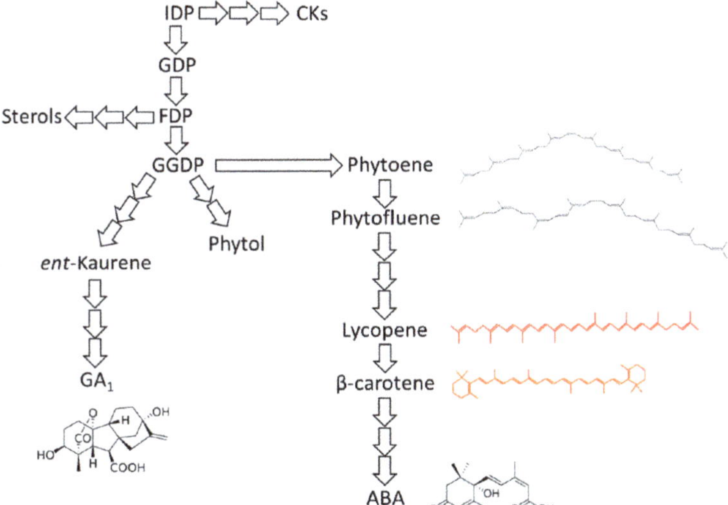

Figure 1. General scheme of pathways leading to production of different hormones. ABA and GAs share their biosynthetic pathway with other hormones, such as CKs. During fruit maturation, balance among hormone biosynthesis changes, and this involves carotenogenesis. IDP, isopentenyl diphosphate; GDP, geranyl diphosphate; FDP, farnesyl diphosphate; GGDP, geranyilgeranyl diphosphate. Number of arrows illustrate number of biosynthetic steps.

The antagonistic effects of both ABA and GAs in regulating different developmental processes and responses to stress are well known; from a biosynthetic point of view, there exists a competition for the metabolic precursor geranylgeranyl pyrophosphate (GGPP) between GA, phytol and carotenoids biosynthetic pathways [34,35]. In the case of GAs,

GGPP undergoes cyclization to *ent*-kaurene and then oxidation to GA_{12}-aldehyde, the precursor of all GAs [36]. ABA is synthesized through C15 intermediates after oxidative cleavage of some xanthophylls [37]. Gibberellins (GAs) are tetracyclic diterpenoid carboxylic acids and are also involved in fruit growth and maturation [38]. More than a hundred GAs have been identified in vascular plants [39], but only a few are biologically active [26,40,41]. The main bioactive forms of GAs are GA_1, GA_3, GA_4 and GA_7. These molecules commonly have a hydroxyl group on C-3β, a carboxyl group on C-6, and a lactone between C-4 and C-10 [40].

3.2. GA Biosynthesis During Fruit Development and Maturation

Previous studies have reported the occurrence of these bioactive forms of GAs in different species including non-climacteric fruits, and differential biosynthesis during the processes of fruit development and maturation. In grape berries, expansion of the berry fruit induced by GA_3 may be linked to the upregulation of cellulose synthase A catalytic subunit genes [42]. Exposure to ABA and GA can induce expression of *Vitis vinifera* Hexose Transporters *VvHT2*, *VvHT3* and *VvHT6* in grape berries during the ripening period when sugar unloading from the phloem is favored [43]. Other genes are also expressed at the late stage of grape berry fruit development and ripening, such as the *Vitis vinifera* SBP-box-like18 (*VvSPL18*), that is significantly upregulated by GA at veraison through an ABA-independent pathway and at the late stage of the berry pericarp ripening process, showing its regulation on maturation [44,45].

In strawberry, recent studies have reported the GA association with fruit development and ripening [41,46–48]. The enlargement of receptacle cells during fruit development is regulated by endogenous GAs [41]. The overexpression of the Gibberellin Stimulated Transcript 2 (*FaGAST2*) gene in different strawberry transgenic lines promoted a reduction in fruit size [46]. Silencing of *FaGAST2* resulted in increase of *FaGAST1* expression, but no changes in fruit cell size were noted; this suggests an orchestrated role of both genes at the transcriptional level in controlling fruit size [26,46].

Accumulation of GAs and their metabolism are also important factors controlling maturation in non-climacteric fruits. The presence of bioactive GA_1, GA_3 and GA_4 has been reported during strawberry fruit development. GA_1 and GA_4 are most abundant at the early stages of fruit development, and decrease as the strawberry fruit ripens [49,50]. The GA_4 content in strawberry receptacles is higher than that of GA_1 and GA_3, which suggests a major role of GA_4 in the developmental processes underlying the receptacle transition from green to white, and subsequently to red [41]. Expression of genes encoding GA pathway components involved in GA biosynthesis (*FaGA3ox*) and catabolism (*FaGA2ox*) is higher in the receptacle during strawberry fruit development [41]. Expression of *FaGA3ox*, which is involved in biosynthesis of bioactive GA_4, is maximum at the green stage while the expression level of *FaGA2ox*, which is involved in the inactivation of this active GA, increases during ripening, and peaking at the red stage [40,41]. Moreover, the expression of *FaGA3ox* in green receptacle is 40 times higher than the expression in the green achene, suggesting that this gene has a prominent role in GA signaling in this tissue [41]. Then, considerable decline in the bioactive forms of GAs is observed at the later stage of fruit development following by the expression of the *FaGA2ox* gene that encodes key enzymes of GA inactivation. These observations, taken together, indicate the degradation of active GA and their content reduction during fruit development, and before maturation processes start.

3.3. Exogenous GA Affect Fruit Maturation and ABA Levels in Fruit

The effect of exogenous GA delaying fruit maturation is well known, and leverage of this knowledge has resulted in common horticultural management practices. For instance, the exogenous application of GA_3 has an inhibitory effect on strawberry ripening, which is evidenced by the delay in anthocyanin synthesis and the decrease in respiration, as well as the reduction in phenylalanine ammonia-lyase (PAL), chlorophyllase and peroxidase activities, enzymes involved in chlorophyll metabolism. This results in delay of degreening [51,52]. Similarly, in citrus GA$_3$ treatment is commonly adopted as a degreening-delay strategy, aimed at managing harvesting dates, when applied before the onset of color break [53]. It has been shown that GA maintains higher content of lutein and prevents accumulation of downstream phytoene, phytofluene and xanthophylls leading to ABA synthesis [54,55].

As mentioned above, ABA and GAs have antagonistic effects in regulating several processes in plants, and their relative balance differs during fruit maturation. In the peel of Navel oranges, concentration of GA$_4$ and GA$_1$ declines before color break and this decline precedes the increase in ABA content. Concentration of both GA and ABA follow then an opposite evolution [16,56]. It has been suggested that the decrease in GA concentration and increase of ABA levels in the peel is part of the ripening program that may stimulate other metabolic pathways associated with coloration, including chlorophyll breakdown and pigment accumulation [57].

4. Integrating Signals to Regulate Maturation: GA, ABA, Sugars, and Ethylene Interaction

Sugar and ABA signaling are closely related in regulating numerous processes in plants and have been studied in detail in the model plant *Arabidopsis thaliana*. Many of these studies are translatable to crop plants of agronomic interest, Genetic studies have identified several loci involved in both sugar and ABA responses, regulating several developmental processes [58]. Interestingly, many sugar-insensitive *Arabidopsis* mutants are either ABA insensitive (*abi* mutants) or ABA deficient (*aba* mutants). There exist many examples of gene co-regulation between sugars and ABA, and in *Arabidopsis*, 14% of genes upregulated by ABA are induced also in response to glucose [59], whereas several other genes involved in stress responses and carbohydrate metabolism are repressed by both regulators [60]. Additionally, there is an increasing body of evidence connecting ABA and sugar signaling during non-climacteric fruit maturation. For instance, in grape berries, a wealth of data correlates increases in sugar and ABA with the onset of ripening [61–64], the ripening-related *ASR* gene is induced by sugar and strongly enhanced by ABA [65], and induction of senescence is ABA-independent, whereas deficiency in the hormone seems to accelerate senescence [66]. Interestingly, in this fruit, synthesis of anthocyanins fails if sugar import into the berry is disrupted via phloem girdling prior to the onset of ripening [64]. In addition, applications of both sugars and ABA, as well as management practices that increase ABA content, also increase anthocyanin accumulation [67–70], and this occurs at the transcriptional level, by induction of gene expression [64]. The genes *VvHT2* and *VvHT6*, that increased expression at veraison after ABA and GA treatment, are the most important sugar transporters across all stages of berry development, with higher expression at the onset of ripening [43]. These authors suggest that both transporters are more related to phloem unloading in sink organs (fruits) than to phloem loading in source organs (leaves). They also emphasize that these transporters may contribute to mobilize a higher content of carbohydrates from leaves to berries, reinforcing the sink strength of fruits at the onset of ripening.

It has been proposed that in fruit from sweet orange, color change during maturation is the consequence of reduction in levels of the active gibberellins GA$_1$ and GA$_4$, involved in the regulation of sugars and ABA accumulation in the rind [56]. In this sense, girdling, a well stablished crop management practice, results in reduction of carbohydrate content and delayed peel coloration, whereas GA levels do not decline in the fruit, indicating the physiological connection among these signals. This also suggests that decrease in GA concentration in the fruit is part of the maturation program, as the presence of gibberellins prevents fruit color change, and that active GA concentration must diminish in fruit to allow color break, whereas increase in ABA content precedes fruit color development [56].

Development and maturation of non-climacteric fruit does not require ethylene biosynthesis. However, many of these fruits respond to ethylene during maturation advancing color or increasing size [26], and sensitivity to ethylene could be the key, playing a pivotal role in the process. It has been proposed that changes in the sensitivity to ethylene may be necessary to maintain coloration in the peel of Citrus fruits and that ABA would enhance sensitivity of the fruit to ethylene, as it has been demonstrated in climacteric fruits [71]. Ethylene would then be the stimulator of transcriptional and biochemical changes ultimately associated with maturation [57]. In this scheme, GA levels would concomitantly be reduced as ABA increased (Figure 2).

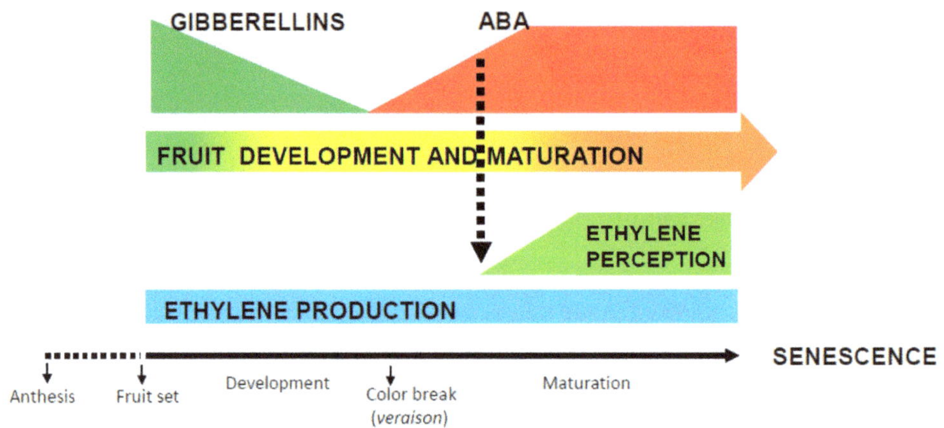

Figure 2. A time-course model of the interplay between abscisic acid (ABA) and gibberellins (GAs) in modulating non-climacteric fruit development and maturation. This is a reductionistic model, as other players involved are not depicted. These include nutritional and environmental factors. The shape of the elements in the figure illustrates the evolution of each component during fruit maturation. GAs decrease, ABA increase and ethylene production remains steady, whereas ethylene perception increases. The crosstalk with ethylene (dash line pointing an induction in ethylene perception driven by ABA) remains to be demonstrated.

Citrus as a Model Plant for Non-Climacteric Maturation Studies

Studies on the role of hormones, their interplay, as well as crosstalk with other factors (i.e., nutritional) in controlling developmental processes and responses to environment have been classically addressed using easy-to-genetically-manipulate model plants. This is the case of *Arabidopsis thaliana*, *Zea mays*, or *Solanum lycopersicum*, in which the availability of mutants impaired in synthesis or perception of a hormone is not a bottleneck for these kind of studies. Although many of the processes studied by using these plant systems can be translated to other agronomically interesting plants, there are specificities, especially in woody plants yielding fleshy fruit that are unique and require more tailored approaches. In many woody plants, usually of agronomic interest, this model plant approach has been traditionally less affordable, due to technical challenges, including in some cases lack of information at the genomic level, unavailability of varieties and/or mutants impaired in hormonal biosynthesis or response, long juvenile period, and difficulties to achieve efficient genetic transformation. Increasingly, this is not the case with Citrus, as in recent years, many species from the genus Citrus have been sequenced, their genealogy revealed, and the sequences made publicly available [72]; Citrus species are prone to spontaneous mutations, with many of these affecting hormonal regulation of maturation, such as 'Pinalate', a spontaneous mutant of Navel orange (*Citrus sinensis* L. Osbeck) that presents lower levels of ABA in all fruit tissues as compared to its parental, and 'Navel negra', a mutant that is impaired in chlorophyll degradation [5,73]; and finally, genetic transformation has been achieved through diverse engineering techniques, and greatly improved with practical, applicable results [74–76]. Together, these advances have helped to elucidate the role of ABA and GAs in the regulation of non-climacteric fruit maturation. The genus Citrus is very diverse, as is comprised by various species and varieties including oranges, mandarins, lemons, grapefruits, pummelos, citrons, limes, kumquats; in addition, different hybrids and spontaneous mutants that have been selected for commercial reasons and are predominantly grown in the tropical and subtropical regions. Citrus develop spontaneous mutations with remarkable frequency in the field. As a result, many of the cultivars currently grown around the world have been obtained by selection of these naturally occurring mutants [5]. Some of these available mutants provide useful aids to dissect some of the processes affected by the mutation [73,77,78]. For instance, the peel of Citrus constitutes an excellent system to investigate the regulation of ABA biosynthesis, signaling and interplay with other hormones and stress regulators during peel maturation [16,79–81]. Recently, it has been completed the identification of ABA signaling core components in Citrus, and their function during maturation has started to be unveiled. This complex is comprised of six PYR/PYL/RCAR ABA receptors, five PP2CAs, and two subclass III SnRK2s. During sweet orange fruit development and ripening, the expression pattern of some ABA receptors mirrors the ABA content, whereas that of *CsPP2CA* genes parallels the hormone accumulation, together modulating ABA perception, downstream signaling, and, consequently, physiological ABA responses [77]. Not only have citrus been useful in understanding fruit maturation though the use of mutants, the response of citrus fruit to different stresses has also started to be elucidated using available mutants defective in ABA, as hormonal signaling in response to stress is also modulated, and varies during maturation [78,81]. This has implications in understanding hormonal regulation of the response to postharvest stress and paves the path to better management practices. In any case, to consider Citrus as a model, the knowledge accrued on these studies should be translatable to other genus.

5. Conclusions

In a nutshell, many studies have been done in the last two decades focusing on the integration of hormonal and nutritional signals during non-climacteric fruit maturation, that has pointed at the interplay between ABA and GAs as a major factor controlling the process. However, the fine details of this regulation are still not well understood and some reports show conflicting results as we have mentioned previously. For instance, how ABA levels may determine tissue sensitivity to ethylene and trigger downstream effects, and how GA and other factors including nutritional and environmental cues, interact in the process, is not completely understood. This warrants future research on how sensitivity to ethylene is triggered and regulated, the involvement of sugars and climate in the whole process, if and how downstream processes depend also on this hormonal setup, if these responses are conserved or species-specific, and—from a practical and commercial standpoint—the implications of this phenomenon during postharvest, as they relate and may determine fruit quality.

Author Contributions: Conceptualization, F.A. writing—original draft preparation, F.A., D.U.d.C. and D.B.; writing—review and editing, F.A. All authors have read and agreed to the published version of the manuscript.

Funding: This research received no external funding.

Conflicts of Interest: The authors declare no conflict of interest.

References

1. Cao, M.; Zheng, J.; Zhao, Y.; Zhang, Z.; Zheng, Z.-L. Network Analysis of Differentially Expressed Genes across Four Sweet Orange Varieties Reveals a Conserved Role of Gibberellin and Ethylene Responses and Transcriptional Regulation in Expanding Citrus Fruits. *Trop. Plant Biol.* **2018**, *12*, 12–20. [CrossRef]
2. Klee, H.J.; Giovannoni, J.J. Genetics and Control of Tomato Fruit Ripening and Quality Attributes. *Annu. Rev. Genet.* **2011**, *45*, 41–59. [CrossRef] [PubMed]

3. Seymour, G.B.; Østergaard, L.; Chapman, N.H.; Knapp, S.; Martin, C. Fruit Development and Ripening. *Annu. Rev. Plant Biol.* **2013**, *64*, 219–241. [CrossRef] [PubMed]
4. Kumar, R.; Khurana, A.; Sharma, A.K. Role of plant hormones and their interplay in development and ripening of fleshy fruits. *J. Exp. Bot.* **2013**, *65*, 4561–4575. [CrossRef]
5. Rodrigo, M.J.; Marcos, J.F.; Alférez, F.; Mallent, M.D.; Zacarías, L. Characterization of Pinalate, a novel *Citrus sinensis* mutant with a fruit-specific alteration that results in yellow pigmentation and decreased ABA content. *J. Exp. Bot.* **2003**, *54*, 727–738. [CrossRef]
6. Fraser, P.D.; Bramley, P. The biosynthesis and nutritional uses of carotenoids. *Prog. Lipid Res.* **2004**, *43*, 228–265. [CrossRef]
7. Cutler, A.J.; Krochko, J.E. Formation and breakdown of ABA. *Trends Plant Sci.* **1999**, *4*, 472–478. [CrossRef]
8. Liotenberg, S.; North, H.; Marion-Poll, A. Molecular biology and regulation of abscisic acid biosynthesis in plants. *Plant Physiol. Biochem.* **1999**, *37*, 341–350. [CrossRef]
9. Taylor, I.B.; Burbidge, A.; Thompson, A.J. Control of abscisic acid synthesis. *J. Exp. Bot.* **2000**, *51*, 1563–1574. [CrossRef]
10. Kondo, S.; Gemma, H. Relationship between Abscisic Acid (ABA) Content and Maturation of the Sweet Cherry. *J. Jpn. Soc. Hortic. Sci.* **1993**, *62*, 63–68. [CrossRef]
11. Ren, J.; Sun, L.; Wu, J.; Zhao, S.; Wang, C.; Wang, Y.; Ji, K.; Leng, P. Cloning and expression analysis of cDNAs for ABA 8′-hydroxylase during sweet cherry fruit maturation and under stress conditions. *J. Plant Physiol.* **2010**, *167*, 1486–1493. [CrossRef] [PubMed]
12. Kondo, S.; Ponrod, W.; Kanlayanarat, S.; Hirai, N. Abscisic Acid Metabolism during Fruit Development and Maturation of Mangosteens. *J. Am. Soc. Hortic. Sci.* **2002**, *127*, 737–741. [CrossRef]
13. Goldschmidt, E.E.; Goren, R.; Even-Chen, Z.; Bittner, S. Increase in Free and Bound Abscisic Acid during Natural and Ethylene-induced Senescence of Citrus Fruit Peel. *Plant Physiol.* **1973**, *51*, 879–882. [CrossRef] [PubMed]
14. Harris, M.J.; Dugger, W.M. The Occurrence of Abscisic Acid and Abscisyl-β-d-Glucopyranoside in Developing and Mature Citrus Fruit as Determined by Enzyme Immunoassay. *Plant Physiol.* **1986**, *82*, 339–345. [CrossRef] [PubMed]
15. Romero, P.; Lafuente, M.T.; Rodrigo, M.J. The Citrus ABA signalosome: Identification and transcriptional regulation during sweet orange fruit ripening and leaf dehydration. *J. Exp. Bot.* **2012**, *63*, 4931–4945. [CrossRef] [PubMed]
16. Alférez, F. Regulación Hormonal de la Maduración en Frutos Cítricos y Su relación Con Alteraciones Fisiológicas Durante la Postcosecha. Ph.D. Thesis, University of Valencia, Valencia, Spain, 2001.
17. Aung, L.H.; Houck, L.G.; Norman, S.M. The Abscisic Acid Content of Citrus with Special Reference to Lemon. *J. Exp. Bot.* **1991**, *42*, 1083–1088. [CrossRef]
18. Valero, D.; Martínez-Romero, D.; Serrano, M.; Riquelme, F. Influence of Postharvest Treatment with Putrescine and Calcium on Endogenous Polyamines, Firmness, and Abscisic Acid in Lemon (*Citrus lemon* L. Burm Cv. Verna). *J. Agric. Food Chem.* **1998**, *46*, 2102–2109. [CrossRef]
19. Lafuente, M.T.; Martinez-Tellez, M.A.; Zacarias, L. Abscisic acid in the response of Fortune mandarin to chilling. Effects of maturity and high temperature conditioning. *J. Sci. Food Agric.* **1997**, *73*, 494–502. [CrossRef]
20. Brisker, H.E.; Goldschmidt, E.E.; Goren, R. Ethylene-induced Formation of ABA in Citrus Peel as Related to Chloroplast Transformations. *Plant Physiol.* **1976**, *58*, 377–379. [CrossRef]
21. Peppi, M.C.; Fidelibus, M.; Dokoozlian, N.; Walker, M.A. Abscisic acid applications improve the color of crimson seedless table grapes. *Am. J. Enol. Vitic.* **2006**, *57*, 388A.
22. Peppi, M.C.; Walker, M.A.; Fidelibus, M.W. Application of abscisic acid rapidly upregulated UFGT gene expression and improved color of grape berries. *Vitis* **2008**, *47*, 11–14.
23. Wheeler, S.; Loveys, B.; Ford, C.; Davies, C. The relationship between the expression of abscisic acid biosynthesis genes, accumulation of abscisic acid and the promotion of *Vitis vinifera* L. berry ripening by abscisic acid. *Aust. J. Grape Wine Res.* **2009**, *15*, 195–204. [CrossRef]
24. Koyama, K.; Sadamatsu, K.; Goto-Yamamoto, N. Abscisic acid stimulated ripening and gene expression in berry skins of the Cabernet Sauvignon grape. *Funct. Integr. Genom.* **2009**, *10*, 367–381. [CrossRef] [PubMed]
25. Sun, L.; Zhang, M.; Ren, J.; Qi, J.; Zhang, G.; Leng, P. Reciprocity between abscisic acid and ethylene at the onset of berry ripening and after harvest. *BMC Plant Biol.* **2010**, *10*, 257. [CrossRef]
26. Fuentes, L.; Figueroa, C.R.; Valdenegro, M. Recent Advances in Hormonal Regulation and Cross-Talk during Non-Climacteric Fruit Development and Ripening. *Horticulturae* **2019**, *5*, 45. [CrossRef]
27. Wang, H.; Huang, H.; Huang, X. Differential effects of abscisic acid and ethylene on the fruit maturation of Litchi chinensis Sonn. *Plant Growth Regul.* **2007**, *52*, 189–198. [CrossRef]
28. Rehman, M.; Singh, Z.; Khurshid, T. Pre-harvest spray application of abscisic acid (S-ABA) regulates fruit colour development and quality in early maturing M7 Navel orange. *Sci. Hortic.* **2018**, *229*, 1–9. [CrossRef]
29. Wang, X.; Yin, W.; Wu, J.; Chai, L.; Yi, H. Effects of exogenous abscisic acid on the expression of citrus fruit ripening-related genes and fruit ripening. *Sci. Hortic.* **2016**, *201*, 175–183. [CrossRef]
30. Iglesias, D.J.; Cercós, M.; Colmenero-Flores, J.M.; Naranjo, M.A.; Ríos, G.; Carrera, E.; Ruiz-Rivero, O.; Lliso, I.; Morillon, R.; Tadeo, F.R.; et al. Physiology of citrus fruiting. *Braz. J. Plant Physiol.* **2007**, *19*, 333–362. [CrossRef]
31. Zhang, L.; Ma, G.; Kato, M.; Yamawaki, K.; Takagi, T.; Kiriiwa, Y.; Ikoma, Y.; Matsumoto, H.; Yoshioka, T.; Nesumi, H. Regulation of carotenoid accumulation and the expression of carotenoid metabolic genes in citrus juice sacs in vitro. *J. Exp. Bot.* **2011**, *63*, 871–886. [CrossRef]

32. Neill, S.J.; Horgan, R.; Parry, A.D. The carotenoid and abscisic acid content of viviparous kernels and seedlings of *Zea mays* L. *Planta* **1986**, *169*, 87–96. [CrossRef] [PubMed]

33. Paiva, R.; Kriz, A. Effect of abscisic acid on embryo-specific gene expression during normal and precocious germination in normal and viviparous maize (*Zea mays*) embryos. *Planta* **1994**, *192*, 332–339. [CrossRef]

34. Fray, R.G.; Grierson, D. Identification and genetic analysis of normal and mutant phytoene synthase genes of tomato by sequencing, complementation and co-suppression. *Plant Mol. Biol.* **1993**, *22*, 589–602. [CrossRef] [PubMed]

35. Fray, R.G.; Wallace, A.; Fraser, P.; Valero, D.; Hedden, P.; Bramley, P.M.; Grierson, D. Constitutive expression of a fruit phytoene synthase gene in transgenic tomatoes causes dwarfism by redirecting metabolites from the gibberellin pathway. *Plant J.* **1995**, *8*, 693–701. [CrossRef]

36. Talon, M.; Koornneef, M.; Zeevaart, J.A. Endogenous gibberellins in Arabidopsis thaliana and possible steps blocked in the biosynthetic pathways of the semidwarf ga4 and ga5 mutants. *Proc. Natl. Acad. Sci. USA* **1990**, *87*, 7983–7987.

37. Milborrow, B.V. The pathway of biosynthesis of abscisic acid in vascular plants: A review of the present state of knowledge of ABA biosynthesis. *J. Exp. Bot.* **2001**, *52*, 1145–1164. [CrossRef]

38. French, E.; Iyer-Pascuzzi, A.S. A role for the gibberellin pathway in biochar-mediated growth promotion. *Sci. Rep.* **2018**, *8*, 5389. [CrossRef]

39. Macmillan, J. Occurrence of Gibberellins in Vascular Plants, Fungi, and Bacteria. *J. Plant Growth Regul.* **2001**, *20*, 387–442. [CrossRef]

40. Yamaguchi, S. Gibberellin Metabolism and its Regulation. *Annu. Rev. Plant Biol.* **2008**, *59*, 225–251. [CrossRef]

41. Csukasi, F.; Osorio, S.; Gutierrez, J.R.; Kitamura, J.; Giavalisco, P.; Nakajima, M.; Fernie, A.R.; Rathjen, J.P.; Botella, M.A.; Valpuesta, V.; et al. Gibberellin biosynthesis and signalling during development of the strawberry receptacle. *New Phytol.* **2011**, *191*, 376–390. [CrossRef]

42. Wang, X.; Zhao, M.; Wu, W.; Korir, N.K.; Qian, Y.; Wang, Z. Comparative transcriptome analysis of berry-sizing effects of gibberellin (GA3) on seedless *Vitis vinifera* L. *Genes Genom.* **2017**, *39*, 493–507. [CrossRef]

43. Murcia, G.; Pontin, M.; Piccoli, P. Role of ABA and Gibberellin A3 on gene expression pattern of sugar transporters and invertases in *Vitis vinifera* cv. Malbec during berry ripening. *Plant Growth Regul.* **2017**, *84*, 275–283. [CrossRef]

44. Fasoli, M.; Santo, S.D.; Zenoni, S.; Tornielli, G.B.; Farina, L.; Zamboni, A.; Porceddu, A.; Venturini, L.; Bicego, M.; Murino, V.; et al. The Grapevine Expression Atlas Reveals a Deep Transcriptome Shift Driving the Entire Plant into a Maturation Program. *Plant Cell* **2012**, *24*, 3489–3505. [CrossRef] [PubMed]

45. Xie, Z.; Su, Z.; Wang, W.; Guan, L.; Bai, Y.; Zhu, X.; Wang, X.; Jia, H.; Fang, J.; Wang, C. Characterization of VvSPL18 and Its Expression in Response to Exogenous Hormones during Grape Berry Development and Ripening. *Cytogenet. Genome Res.* **2019**, *159*, 97–108. [CrossRef]

46. Moyano-Cañete, E.; Bellido, M.L.; García-Caparrós, N.; Medina-Puche, L.; Amil-Ruiz, F.; González-Reyes, J.A.; Caballero, J.L.; Garciía-Limones, C.; Blanco-Portales, R. FaGAST2, a Strawberry Ripening-Related Gene, Acts Together with FaGAST1 to Determine Cell Size of the Fruit Receptacle. *Plant Cell Physiol.* **2012**, *54*, 218–236. [CrossRef]

47. An, L.; Ma, J.; Wang, H.; Li, F.; Qin, D.; Wu, J.; Zhu, G.; Zhang, J.; Yuan, Y.; Zhou, L.; et al. NMR-based global metabolomics approach to decipher the metabolic effects of three plant growth regulators on strawberry maturation. *Food Chem.* **2018**, *269*, 559–566. [CrossRef]

48. Gu, T.; Jia, S.; Huang, X.; Wang, L.; Fu, W.; Huo, G.; Gan, L.; Ding, J.; Li, Y. Transcriptome and hormone analyses provide insights into hormonal regulation in strawberry ripening. *Planta* **2019**, *250*, 145–162. [CrossRef]

49. Kim, J.; Lee, J.G.; Hong, Y.; Lee, E.J. Analysis of eight phytohormone concentrations, expression levels of ABA biosynthesis genes, and ripening-related transcription factors during fruit development in strawberry. *J. Plant Physiol.* **2019**, *239*, 52–60. [CrossRef]

50. Symons, G.M.; Chua, Y.-J.; Ross, J.J.; Quittenden, L.J.; Davies, N.W.; Reid, J.B. Hormonal changes during non-climacteric ripening in strawberry. *J. Exp. Bot.* **2012**, *63*, 4741–4750. [CrossRef]

51. Martinez, G.A.; Chaves, A.R.; Añon, M.C. Effect of gibberellic acid on ripening of strawberry fruits (*Fragaria annanassa* Duch.). *J. Plant Growth Regul.* **1994**, *13*, 87–91. [CrossRef]

52. Martinez, G.A.; Chaves, A.R.; Añon, M.C. Effect of exogenous application of gibberellic acid on color change and phenylalanine ammonia-lyase, chlorophyllase, and peroxidase activities during ripening of strawberry fruit (*Fragaria* x *ananassa* Duch.). *J. Plant Growth Regul.* **1996**, *15*, 139–146. [CrossRef]

53. García-Luis, A.; Herrero-Villén, A.; Guardiola, J. Effects of applications of gibberellic acid on late growth, maturation and pigmentation of the Clementine mandarin. *Sci. Hortic.* **1992**, *49*, 71–82. [CrossRef]

54. Alós, E.; Cercós, M.; Rodrigo, M.J.; Zacarías, L.; Talon, M. Regulation of Color Break in Citrus Fruits. Changes in Pigment Profiling and Gene Expression Induced by Gibberellins and Nitrate, Two Ripening Retardants. *J. Agric. Food Chem.* **2006**, *54*, 4888–4895. [CrossRef] [PubMed]

55. Rodrigo, M.J.; Zacarias, L. Effect of postharvest ethylene treatment on carotenoid accumulation and the expression of carotenoid biosynthetic genes in the flavedo of orange (*Citrus sinensis* L. Osbeck) fruit. *Postharvest Biol. Technol.* **2007**, *43*, 14–22. [CrossRef]

56. Gambetta, G.; Martínez-Fuentes, A.; Bentancur, O.; Mesejo, C.; Reig, C.; Gravina, A.; Agustí, M. Hormonal and nutritional changes in the flavedo regulatingrind color development in sweet orange [*Citrus sinensis* (L.) Osb.]. *J. Plant Growth Regul.* **2012**, *31*, 273–282. [CrossRef]

57. Rodrigo, M.J.; Alquézar, B.; Alós, E.; Lado, J.; Zacarías, L. Biochemical bases and molecular regulation of pigmentation in the peel of Citrus fruit. *Sci. Hortic.* **2013**, *163*, 46–62. [CrossRef]

58. Finkelstein, R.; Gibson, S.I. ABA and sugar interactions regulating development: Cross-talk or voices in a crowd? *Curr. Opin. Plant Biol.* **2002**, *5*, 26–32. [CrossRef]

59. Li, Y.; Lee, K.K.; Walsh, S.; Smith, C.; Hadingham, S.; Sorefan, K.; Cawley, G.C.; Bevan, M. Establishing glucose- and ABA-regulated transcription networks in Arabidopsis by microarray analysis and promoter classification using a Relevance Vector Machine. *Genome Res.* **2006**, *16*, 414–427. [CrossRef] [PubMed]

60. Dekkers, B.J.W.; Schuurmans, J.A.M.J.; Smeekens, S.C.M. Interaction between sugar and abscisic acid signalling during early seedling development in Arabidopsis. *Plant Mol. Biol.* **2008**, *67*, 151–167. [CrossRef] [PubMed]

61. Davies, C.; Boss, P.K.; Robinson, S.P. Treatment of grape berries, a nonclimacteric fruit with a synthetic auxin, retards ripening and alters the expression of developmentally regulated genes. *Plant Physiol.* **1997**, *115*, 1155–1161. [CrossRef] [PubMed]

62. Deluc, L.G.; Quilici, D.R.; Decendit, A.; Grimplet, J.; Wheatley, M.D.; Schlauch, K.; Mérillon, J.-M.; Cushman, J.C.; Cramer, G.R. Water deficit alters differentially metabolic pathways affecting important flavor and quality traits in grape berries of Cabernet Sauvignon and Chardonnay. *BMC Genom.* **2009**, *10*, 212–233. [CrossRef] [PubMed]

63. Mei, Z.; Ping, L.; Guanglian, Z.; Xiangxin, L. Cloning and functional analysis of 9-cis-epoxycarotenoid dioxygenase (NCED) genes encoding a key enzyme during abscisic acid biosynthesis from peach and grape fruits. *J. Plant Physiol.* **2009**, *166*, 1241–1252.

64. Gambetta, G.A.; Matthews, M.A.; Shaghasi, T.H.; McElrone, A.J.; Castellarin, S.D. Sugar and abscisic acid signaling orthologs are activated at the onset of ripening in grape. *Planta* **2010**, *232*, 219–234. [CrossRef] [PubMed]

65. Çakir, B.; Agasse, A.; Gaillard, C.; Saumonneau, A.; Delrot, S.; Atanassova, R. A Grape ASR Protein Involved in Sugar and Abscisic Acid Signaling. *Plant Cell* **2003**, *15*, 2165–2180. [CrossRef]

66. Pourtau, N.; Mares, M.; Purdy, S.; Quentin, N.; Ruël, A.; Wingler, A. Interactions of abscisic acid and sugar signaling in the regulation of leaf senescence. *Planta* **2004**, *219*, 765–772. [CrossRef]

67. Hiratsuka, S.; Onodera, H.; Kawai, Y.; Kubo, T.; Itoh, H.; Wada, R. ABA and sugar effects on anthocyanin formation in grape berry cultured in vitro. *Sci. Hortic.* **2001**, *90*, 121–130. [CrossRef]

68. Larronde, F.; Krisa, S.; Decendit, A.; Chèze, C.; Deffieux, G.; Mérillon, J.-M. Regulation of polyphenol production in *Vitis vinifera* cell suspension cultures by sugars. *Plant Cell Rep.* **1998**, *17*, 946–950. [CrossRef]

69. Matsushima, J.; Hiratsuka, S.; Taniguchi, N.; Wada, R.; Suzaki, N. Anthocyanin accumulation and sugar content in the skin of grape cultivar Olympia treated with ABA. *J. Jpn. Soc. Hortic. Sci.* **1989**, *58*, 551–555. [CrossRef]

70. Pirie, A.; Mullins, M.G. Changes in Anthocyanin and Phenolics Content of Grapevine Leaf and Fruit Tissues Treated with Sucrose, Nitrate, and Abscisic Acid. *Plant Physiol.* **1976**, *58*, 468–472. [CrossRef]

71. Eveland, A.L.; Jackson, D. Sugars, signalling, and plant development. *J. Exp. Bot.* **2012**, *63*, 3367–3377. [CrossRef]

72. Wu, G.A.; Terol, J.; Ibanez, V.; López-García, A.; Pérez-Román, E.; Borredá, C.; Domingo, C.; Tadeo, F.R.; Carbonell-Caballero, J.; Alonso, R.; et al. Genomics of the origin and evolution of Citrus. *Nat. Cell Biol.* **2018**, *554*, 311–316. [CrossRef] [PubMed]

73. Alós, E.; Roca, M.; Iglesias, D.J.; Minguez-Mosquera, M.I.; Damasceno, C.M.; Thannhauser, T.W.; Rose, J.K.; Talon, M.; Cercos, M. An evaluation of the basis and consequences of a stay-green mutation in the navel negra (nan) citrus mutant using transcriptomic and proteomic profiling and metabolite analysis. *Plant Physiol.* **2008**, *147*, 1300–1315. [CrossRef] [PubMed]

74. Peña, L.; Martín-Trillo, M.; Juárez, J.; Pina, J.A.; Navarro, L.; Martínez-Zapater, J.M. Constitutive expression of Arabidopsis LEAFY or APETALA1 genes in citrus reduces their generation time. *Nat. Biotechnol.* **2001**, *19*, 263–267. [CrossRef] [PubMed]

75. Jia, H.; Orbovic, V.; Jones, J.B.; Wang, N. Modification of the PthA4 effector binding elements in Type I CsLOB1 promoter using Cas9/sgRNA to produce transgenic Duncan grapefruit alleviating XccDpthA4:dCsLOB1.3 infection. *Plant Biotechnol. J.* **2016**, *14*, 1291–1301. [CrossRef]

76. Dutt, M.; Barthe, G.; Irey, M.; Grosser, J. Transgenic Citrus Expressing an Arabidopsis NPR1 Gene Exhibit Enhanced Resistance against Huanglongbing (HLB.; Citrus Greening). *PLoS ONE* **2015**, *10*, e0137134. [CrossRef]

77. Romero, P.; Rodrigo, M.J.; Alférez, F.; Ballester, A.-R.; González-Candelas, L.; Zacarías, L.; Lafuente, M.T. Unravelling molecular responses to moderate dehydration in harvested fruit of sweet orange (*Citrus sinensis* L. Osbeck) using a fruit-specific ABA-deficient mutant. *J. Exp. Bot.* **2012**, *63*, 2753–2767. [CrossRef]

78. Romero, P.; Lafuente, M.T.; Alferez, F. A transcriptional approach to unravel the connection between phospholipases A2 and D and ABA signal in citrus under water stress. *Plant Physiol. Biochem.* **2014**, *80*, 23–32. [CrossRef]

79. Rodrigo, M.J.; Alquézar, B.; Zacarías, L. Cloning and characterization of two9-cis-epoxycarotenoid dioxygenase genes, differentially regulated during fruitmaturation and under stress conditions, from orange (*Citrus sinensis* L. Osbeck). *J. Exp. Bot.* **2006**, *57*, 633–643. [CrossRef]

80. Alferez, F.; Zacarias, L. Interaction between ethylene and abscisic acid in the regulation of citrus fruit maturation. In *Biology and Biotechnology of the Plant Hormone Ethylene II*; Kanellis, A.K., Chang, C., Klee, H., Bleecker, A.B., Pech, J.C., Grierson, D., Eds.; Kluwer Academic Publishers: Doordrecht, The Netherlands, 1999.

81. Romero, P.; Gandía, M.; Alferez, F. Interplay between ABA and phospholipases A2 and D in the response of citrus fruit to postharvest dehydration. *Plant Physiol. Biochem.* **2013**, *70*, 287–294. [CrossRef]

International Journal of
Molecular Sciences

Review

Abscisic Acid Mediates Drought and Salt Stress Responses in *Vitis vinifera*—A Review

Daniel Marusig and Sergio Tombesi *

Dipartimento di Scienze delle Produzioni Vegetali Sostenibili, Università Cattolica del Sacro Cuore, 29122 Piacenza, Italy; daniel.marusig@unicatt.it
* Correspondence: sergio.tombesi@unicatt.it; Tel.: +39-0523-599221

Received: 21 October 2020; Accepted: 15 November 2020; Published: 17 November 2020

Abstract: The foreseen increase in evaporative demand and reduction in rainfall occurrence are expected to stress the abiotic constrains of drought and salt concentration in soil. The intensification of abiotic stresses coupled with the progressive depletion in water pools is a major concern especially in viticulture, as most vineyards rely on water provided by rainfall. Because its economical relevance and its use as a model species for the study of abiotic stress effect on perennial plants, a significant amount of literature has focused on *Vitis vinifera*, assessing the physiological mechanisms occurring under stress. Despite the complexity of the stress-resistance strategy of grapevine, the ensemble of phenomena involved seems to be regulated by the key hormone abscisic acid (ABA). This review aims at summarizing our knowledge on the role of ABA in mediating mechanisms whereby grapevine copes with abiotic stresses and to highlight aspects that deserve more attention in future research.

Keywords: ABA; grapevine; stomata; drought; metabolism; carbohydrates; salinity

---

1. Introduction

Climate change is expected to have negative impacts on the socioeconomic system [1]. Despite the discrepancy between the projected scenarios, even the most optimistic models foresee an increase in the occurrence and duration of anomalous droughts, especially in the Mediterranean-climate regions, where water sources will be increasingly scarce [2]. These threats are of major concern to agriculture and in particular to viticulture, being the one of the most profitable crops in these regions [3]. In 2016, less than 10% of vineyards in the European Union were irrigated, even if they accounted for approximatively 60% of world's grape production [3]. Thanks to the lately moderation of restrictions imposed by law, in many states the use of irrigation has increased (e.g., in 2018, the irrigated vineyards' area in Spain exceeded 30%) [4,5]. The deleterious effects of climate change will negatively affect water sources extent and quality, and the solely irrigation is not a sustainable and sufficient strategy to counteract the expected impacts on grape production [3]. In order to cope with these constraints, a general improvement in viticulture techniques is needed and the achievement of this goal requires a better knowledge of grapevine physiology under stress conditions.

Vitis vinifera is a Mediterranean vine [6] mostly cultivated in Mediterranean-like areas, and in particularly arid environments like Karst [7]. Grapevine is adapted to cope with drought conditions and some common cultivation techniques are based on imposing moderate soil water deficits, in order to improve the quality of berries, minimize the yield reduction and favor the production of flavonoids, sugars, polyphenols, and carotenoids [5].

Further to its economical relevance, *V. vinifera* is a model species in drought-response investigation [8]. The drought-tolerance strategy of grapevine consists in an ensemble of interactions between morphological/structural traits and a pronounced control of water loss by stomatal regulation [5].

The latter one is mostly mediated by hormonal regulation, and a pivotal role in this process is played by the abscisic acid (ABA) [9].

ABA is a ubiquitous hormone, which has been discovered to modulate physiological responses among all the kingdoms of life [10]. In vascular plants, it regulates a multitude of physiological processes like seed and bud dormancy [11], cambium activity [12], organs development [13], and fruit ripening [14]. Yet, it is involved in mediating physiological responses to stressful conditions, especially excessive temperature, salinity, and drought [15–18]. ABA has been mainly investigated as it promotes the loss of turgor pressure in guard cells triggering stomatal closure [9], however, an increasing amount of evidences, also based on *V. vinifera*, address to ABA a central role in other fundamental processes as the hydraulic response to prolonged drought conditions [19], the regulation of carbohydrate metabolism during recovery [20], and the exclusion of excessive salts dissolved in the soil solution [21].

Since the socioeconomic importance of *V. vinifera* and the increasing number of studies asserting ABA to be a major factor in regulating a so far underestimated amount of physiological responses, the aims of this review are: (i) to provide an exhaustive overview on the role of ABA in *V. vinifera* as a key hormone involved in regulating the mechanisms for coping with the major threats caused by climate change and (ii) highlight the main issues that deserve further investigation.

2. ABA Biosynthesis and Translocation

ABA is a 15-carbons isoprenoid, derived from the metabolism of β-carotene (40 carbons), originated from the methylerythritol 4-phosphate (MEP) pathway [22]. The β-carotene to ABA-biosynthetic pathway is composed by many steps, and it is schematized in Figure 1. In the first part of the pathway (40-carbons), the β-carotene is converted into the isomer zeaxanthin (Zx), which is metabolized through an ensemble of conversions called the "xanthophyll cycle," having final product violaxanthin (Vx) and neoxanthin (Nx). Both Vx and Nx can be converted into the respective 9-cis-isomers, from which cleavage, the sequisterpenoid xanthoxin (Xx) is produced. At this step, the 15-carbons pathway starts leading to the synthesis of ABA [9].

Being ABA produced by cleavage of carotenoids, research has initially focused on investigating the leaf as main site of its biosynthesis [23–25]. In 1974, Loveys and Kriedemann [23] investigated the role of leaf-produced-ABA causing stomatal closure in *V. vinifera* cv. Cabernet Sauvignon, by measuring ABA concentration ([ABA]) in detached leaves. Their results assessed significant increases in [ABA] and stomatal resistance, supporting the hypothesis of endogenous ABA triggering stomatal closure. A few years later, the same authors [24] measured [ABA] in extracts of *Spinacea oleracea* leaves, speculating that most of the ABA is contained in the chloroplasts, and it is present also in nonstressed leaves. In contrast to these results, evidences on *Vitis vinifera* cv. Riesling and Silvaner pointed out that [ABA] in xylem sap ($[ABA]_{xy}$) was related to stomatal conductance (g_s) regulation, suggesting that ABA may be translocated to the roots through the phloem and back to the shoot through the xylem, guaranteeing a continuous supply to leaves [25].

Since ABA biosynthesis in the leaf would require stress-induced stimuli (e.g., loss of turgor and shrinkage) [26], root system was proposed as a primary site for ABA biosynthesis [27,28]. In 1987, Zhang and Davies [27] observed the production of ABA in detached root tips of *Pisum sativum* and *Commelina communis*, providing one of the first strong evidences supporting the hypothesis that ABA is synthesized in root tips and transported from roots to shoot via the transpiration stream. Soar et al. [29] reported that in Shiraz vines, $[ABA]_{xy}$ increased as drought was more intense. The increase in $[ABA]_{xy}$ changed according to the different rootstock, suggesting that the different root system may affect the ABA production and delivery [29]. These conclusions have been supported also by gene-expression analyses. In 2013, Speirs et al. [30] measured [ABA] and expression of the ABA biosynthesis genes VviNCED1 and VviNCED2, in Cabernet Sauvignon grapevine roots and leaves. As drought stress increased, [ABA] strongly increased in roots, xylem, and leaves, especially in the latter. Nevertheless, gene expression remained stable in leaves, but remarkably increased in roots.

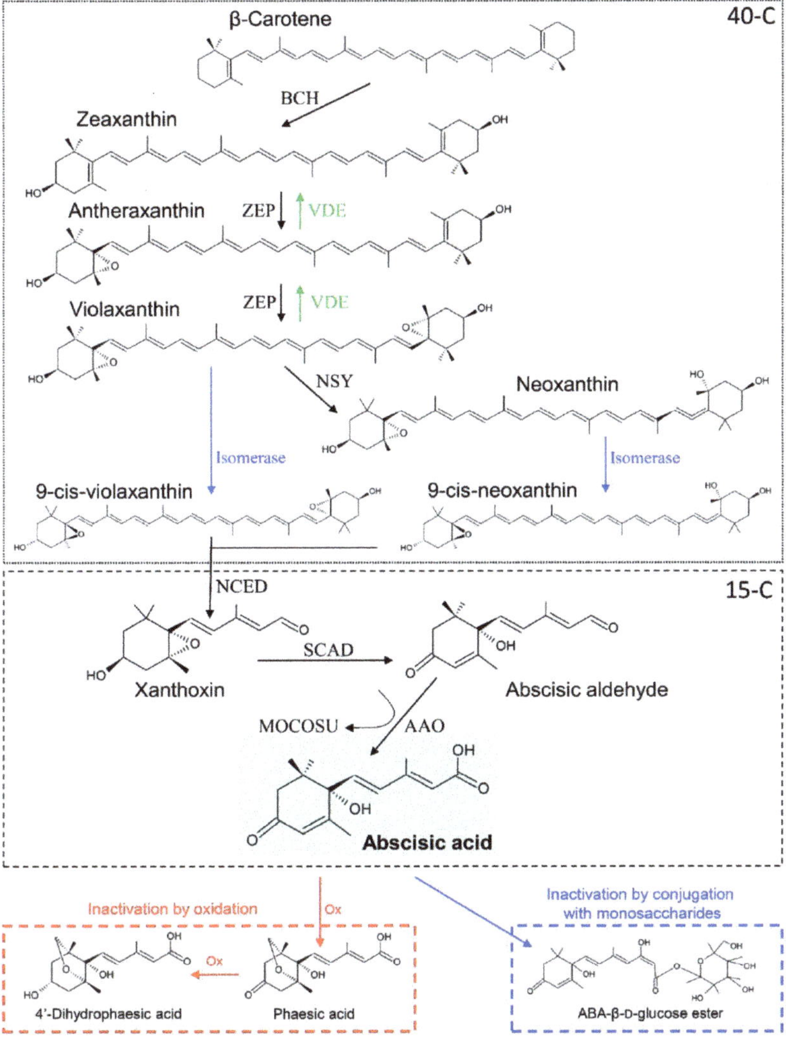

Figure 1. Abscisic acid biosynthesis and metabolism. In the first step, β-carotene is di-hydroxylated by β-carotene hydroxilase (BCH) proteins to produce the transisomer zeaxanthin. Hence, zeaxanthin is epoxidated by zeaxanthin oxidase (ZEP) to antheraxanthin and, then, to violaxanthin. ZEP-mediated reactions can be reversed by violaxanthin de-epoxidase (VDE). Violaxanthin can be transformed into neoxanthin by neoxanthin synthase (NSY) and both violaxanthin and neoxanthin are converted in the respective 9-cis-isomer by isomerase catalysts. The 15-carbons apocarotenoid sesquiterpenoid xanthoxin is then produced by cis-xanthophylls cleavage, whose reaction is catalyzed by 9-cis-epoxycarotenoid dioxygenase (NCED). Subsequently, xanthoxin is oxidized to abscisic aldehyde by short-chain alcohol dehydrogenase (SCAD), and finally abscisic acid (ABA) is produced by oxidation of abscisic aldehyde through the combined action of ABA-aldehyde oxidase (AAO) and a molybdenum cofactor sulfurase (MOCOSU). ABA can be inactivated by oxidation or by conjugation with monosaccharides. In the first way, ABA is oxidized (Ox) at first to phaseic acid and then to 4'-dihydrophaseic acid. In the second one, ABA is conjugated with glucose to produce ABA-β-D-glucose ester. Based on [9].

These findings led to the development of an irrigation technique, named "Partial Root Drying (PRD)," aimed at optimizing grapevine water use, in order to reduce irrigation and improve berry quality [31–34]. This objective is pursued inducing ABA production in the root system, hence limiting water loss by triggering stomatal closure [31]. From a technical point of view, while water supply is provided by watering part of the root system, in the remaining part, irrigation is withdrawn, in order to stimulate ABA production [35]. Hence, ABA is transported through the xylem to the leaves by the driving force generated from transpiration (see Section 3) [17]. In order to guarantee a stable production and translocation of ABA, the wet and dry parts of the roots are weekly switched [17,36]. In 2007, Poni et al. [32] evaluated the physiological response and the yield-quality performance of Sangiovese grapevines under PRD irrigation regime. Their results highlighted that under PRD, the water use efficiency was improved as g_s was strongly reduced, while the decrease in photosynthetic assimilation (A) was more limited. If compared to the well-watered (WW) treatment, the PRD had also positive effects on the control of vigor, it improved the berry quality and it did not cause variations in yield [32]. Romero et al. [34] investigated physiological responses of Monastrell grapevines under different irrigation regimes, for 4 years. Under PRD, $[ABA]_{xy}$ significantly increased, following the trend of depletion in soil water availability. Moreover, under PRD, vines developed a deeper root system and maintained a better water status than those irrigated with the same water volumes but with regulated deficit irrigation (RDI), where water was totally withdrawn for a limited period [34]. Even an RDI strategy would set a portion of the root system that insists into a dry soil volume and stimulate ABA production. However, $[ABA]_{xy}$ seems to depend on the volumetric soil water content of both wet and dry sides [16].

The hypothesis of ABA being mainly synthetized in the root system has been commonly accepted [37], and it is still supported by more recent research [38]. Despite that, an overwhelming body of literature has been supporting the hypothesis that drought-induced stomatal closure is also mediated by in-site produced ABA into the leaf [39–44]. A first issue is represented by precursors abundancy: the ABA biosynthetic pathway is based on carotenoid, whose accumulation depends on the ability of the plastid to sequester them in specific sinks [45]. Due to the presence of chloroplasts and chromoplasts, this is ordinary in leaves. On the contrary, in roots plastids are mainly proplastids and leucoplasts, which cannot accumulate carotenoids [46], and a root-to-shoot ABA supply pathway seems unlikely [41]. In 2009, Ikegami et al. [40] observed rising [ABA] in roots and leaves of *Arabidopsis thaliana* under drought. However, when the same measurements were replicated on the same tissues detached from the plant, [ABA] increased only in leaves. These observations have been strongly supported also by gene expression analyses [39,47] and solid evidences claim the mesophyll being the main site of ABA biosynthesis in leaf [42,43]. In 2006, Soar et al. [39] measured the diurnal variation of $[ABA]_{xy}$ and expression of ABA-biosynthetic genes in Grenache and Shiraz grapevines. They observed that at midday, under high evaporative demand, [ABA] and the expression of VviNCED1 and VviZEP increased in leaf and remained stable in roots [39]. McAdam and Brodribb [42] measured [ABA] in bench-dried leaves of five different species (one angiosperm and four gymnosperms), which leaf anatomy allowed to isolate mesophyll, vascular tissue, and stomata. They observed that drought-induced production of ABA mainly occurs in the mesophyll [42], hence supporting the hypothesis that mesophyll shrinkage, due to cell volume decline, may trigger the whole process [43].

This topic is still debated, but a few researches aimed at investigating leaf as the biosynthetic source of ABA in grapevine. Indeed, in grapevine, the investigation of stress response was more focused on cultivar [48,49], scion–rootstock interaction [50], observational scale [51], geographical area [49], acclimation, and experienced drought stress [20]. However, the localization of the response to stress mediated by ABA represent an important information for the application of specific viticulture techniques such as irrigation.

3. ABA Role in Regulating Stomatal Closure

Water uptake and transport through the xylem conduits are regulated by a tension-cohesion mechanism [52]. Water loss at the leaf level leads to a decrease in leaf water potential (Ψ_{leaf}). Hence a dynamic water potential gradient is established, which triggers water flow from roots to leaves [52]. As drought persists and water availability accordingly decreases, tension in the xylem conduits increases and the water transport is weighted down by the occurrence of embolism events [53]. Without restoration, the presence of emboli in the xylem conduits breaks the continuity of the water column and prevents water transport through the xylem, leading in the worst cases to hydraulic failure and plant death [54].

Plants are able to modulate water loss through stomatal closure, a process influenced by an ensemble of environmental (e.g., light intensity), hydraulic (e.g., hydraulic conductance), and endogenous (e.g., hormones) factors [55]. The most important factor between the endogenous ones is ABA, which triggers stomatal closure by inducing loss of turgor in guard cells and prevents stomata premature reopening (Figure 2) [56]. When ABA binds to receptors on the guard cells membrane, it triggers the accumulation in cytosol of Ca^{2+} and the efflux of Cl^- and K^+ ions [57]. Osmolytes displacement causes plasma membrane depolarization and the overall reduction in ion content triggers water efflux by osmosis, causing loss of turgor and stomatal closure [9]. This status persists as long as ABA gets inactivated by oxidation or conjugation with monosaccharides (Figure 1). Then, Ca^{2+} concentration ($[Ca^{2+}]$) decreases, and the osmotic balance gets restored by K^+ uptake [56,57].

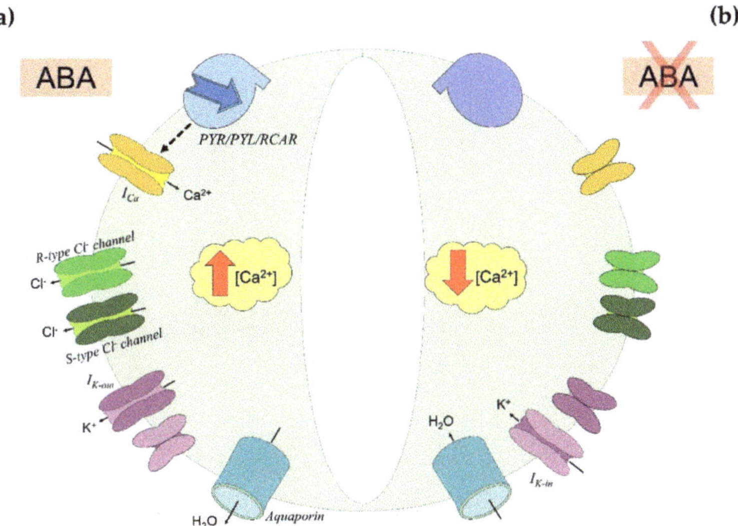

Figure 2. Abscisic acid (ABA) signaling in guard cells. (**a**) ABA inducing stomatal closure. ABA binds to PYR/PYL/RCAR receptors on the guard cells membrane and triggers the accumulation in cytosol of Ca^{2+} by activation of Ca^{2+} channels (I_{Ca}). Under elevated Ca^{2+} concentration ($[Ca^{2+}]$), the cell-efflux of Cl^- is enhanced. This efflux is mediated by rapid transient (R-type) and slow-activating sustained (S-type) Cl^- channels, and it causes plasma membrane depolarization. Thence, the K^+ uptake is downregulated by inward-rectifying K^+ channels ($I_{k\text{-}in}$) activity, while the K^+ efflux is promoted through outward-rectifying K^+ channels ($I_{k\text{-}out}$). The overall reduction in ions content triggers water efflux through aquaporins by osmosis, causing loss of turgor in guard cells and stomatal closure. (**b**) Stomatal reopening by ABA inactivation. As ABA does not bind further to PYR/PYL/RCAR receptors, Ca^{2+} accumulation ceases. The osmotic balance is restored by K^+ uptake through $I_{k\text{-}in}$, promoting water uptake and reacquiring turgidity. Based on [56].

ABA role in stomatal regulation has been particularly studied in grapevine, in order to optimize irrigation techniques and genotype-specific response to drought. Irrigation techniques aim at maintaining plant water status and at maximizing plant water use efficiency (WUE), expressed as the ratio between photosynthetic CO_2 assimilation (A) and water loss by transpiration (E) [48]. In grapevine, cultivars differently regulate water loss as they differ in terms of stomatal response under high evaporative demand [5]. This behavior has been addressed using the concept of iso-/anisohydry [58] that generally refers to the daily variation of the difference between Ψ_{leaf} and soil water potential (Ψ_{soil}). Specifically, isohydric cultivars maintain a more constant water status through preventive stomata closure, while anisohydric cultivars keep stomata open until more negative Ψ_{leaf} values [59]. Isohydry has been associated to differences in the ABA signaling process [58]. Since *V. vinifera* displayed a high intraspecific plasticity in anisohydric behavior, it became one of the most investigated species (Figure 3) [60]. In 2001, Bota et al. [48] investigated gas exchange and hydraulic responses to drought in 22 grapevine cultivars. Despite most of the cultivars belonged to the same region, as drought persisted, they displayed a remarkably divergent iso-/anisohydric behavior. Differences in anisohydricity have also been observed by Coupel-Ledru et al. [61], which assessed a divergent stomatal response to drought, even in pseudo-F_1 progeny of a reciprocal cross between the Syrah and Grenache cultivars.

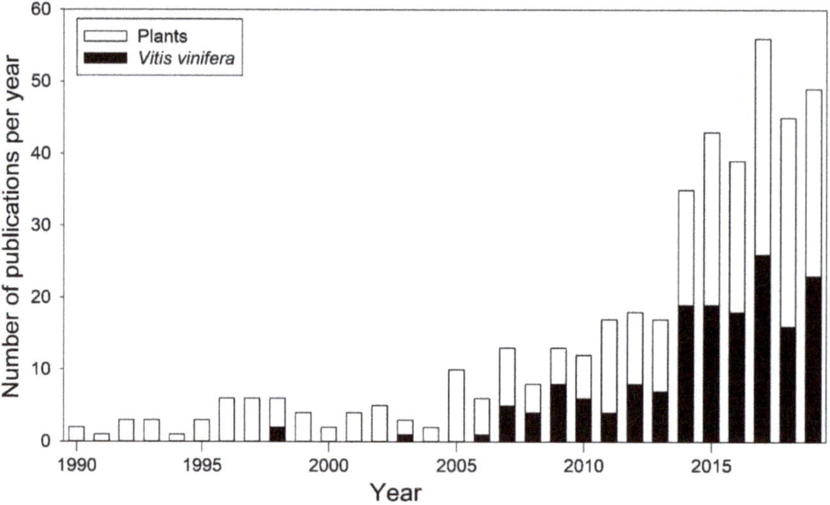

Figure 3. Number of publications dealing with iso-/anisohydry from 1990 to 2019, for all plant species (white bars) and *Vitis vinifera* (black bars) only. Data collected from the Scopus database (https://www.scopus.com/), according to the method of [60].

The intraspecific divergence in stomatal behavior, may be related to a genetic difference in ABA production [58]. Nevertheless, an increasing amount of evidences has highlighted that many other factors are involved in stomatal regulation [5]. Indeed, studies speculating of ABA being the pivotal driver of drought-response have mostly investigated grapevine physiology under severe water stress. However, differences in isohydric behavior can be acknowledged at moderate water stress [49]. Recently, Levin et al. [49] characterized the response of g_s along a wide range of Ψ_{leaf} in 17 grapevine cultivars; they showed that cultivars differed in g_s response only under moderate stress, while under well-watered conditions and severe stress, the intraspecific difference is not appreciable.

As abovementioned, under water shortage, tension in xylem increases inducing the occurrence of embolism events, occluding the conduits and reducing the hydraulic conductivity (K_{xy}) [54]. It has been suggested that cavitation occurrence may act as a hydraulic signal, triggering the stomatal response [62]. This mechanism has been observed on crops [19], herbs [63], and trees [64], also being common between species with contrasting origin and anatomy [62]. Based on this hypothesis, stomatal control strategy of plants should depend on its vulnerability, generally expressed as the water potential at 50% loss of conductivity. This speculation has been strongly supported on a wide range of species [65] and in grapevine [66], however, the debate is still open as some evidences support a more complex interaction [67], also claiming that stomatal closure may precede emboli formation [68].

McAdam and Brodribb [69], in a study on *Metasequoia glyptostroboides*, suggested that stomatal closure may be triggered by passive hydraulic signals, while active ABA-mediated regulation occurs under long-term drought. In grapevine, Tombesi et al. [19] observed that foliar [ABA] significantly increased after g_s was already low, suggesting that stomata closure is regulated differently according to the intensity of drought (Figure 4). As water shortage begins, stomatal closure is likely triggered by hydraulic signals, while long-term ABA maintains turgor loss in guard cells, preventing premature reopening [19]. Similar results have been recently obtained on Merlot grapevines by Degu et al. [70] and may support the evidences of isohydric and anisohydric cultivars displaying different ABA/hydraulic signals stomatal regulation [71].

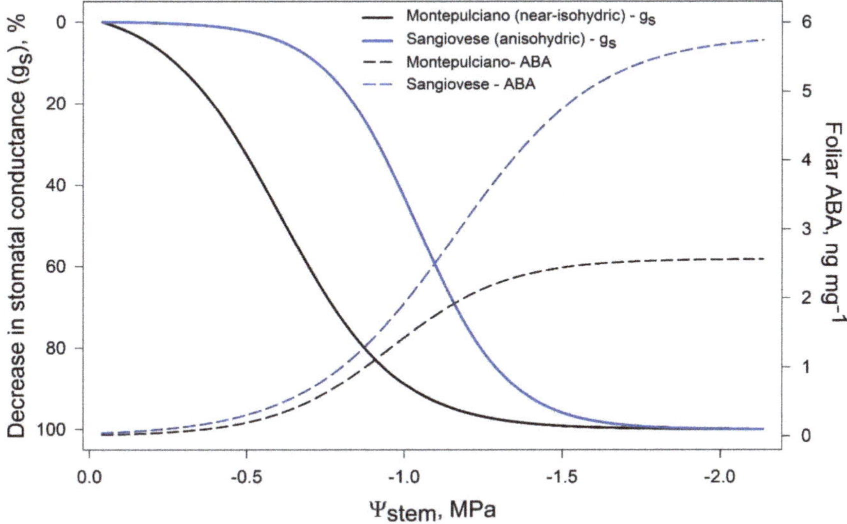

Figure 4. Modelled dynamic of stomatal conductance (g_s) and foliar ABA concentration in a near-isohydric cv (Montepulciano) and an anisohydric cv (Sangiovese). Elaboration on data by [19].

Interestingly, some studies assessed a relation between ABA and hydraulic properties, suggesting that ABA may have an indirect effect on water loss management [72,73]. Soar et al. [39] observed that differences in stomatal responses between Grenache (near-isohydric) and Shiraz (near-anisohydric) were related to [ABA]$_{xy}$. They suggest that the degree of isohydricity may be related to differences in hydraulic properties, which depend on the interaction between ABA and other factors, in particular, aquaporins. In this way, high [ABA] inhibits aquaporins expression, downregulating K_{xy} and inducing stomatal closure [74]. Recently, Dayer et al. [73] investigated the coordination between ABA, aquaporins expression, and hydraulics of root and shoot in Grenache (near-isohydric) and Syrah (near-anisohydric) grapevines under mild water stress. Their results highlighted that even under mild water stress, ABA production causes downregulation of aquaporins expression in the leaf, in order to prevent water loss. The entity of this phenomenon was different according to the cultivar, suggesting that less anisohydric cultivars are more sensible to ABA [73]. Despite these evidences, our knowledge on this topic is still scarce, especially on grapevine [75]. Future research is still needed, in order to better understand the complex relations occurring between ABA and hydraulic regulation, which occurs with a high intraspecific variability.

4. ABA Role in Carbohydrates Mobilization

Threats related to the foreseen increase in occurrence and duration of anomalous dry conditions due to climate change are expected to affect the vegetation in multiple ways. As far as drought persists, plants are forced to keep stomata closed for prolonged periods, limiting the carbon uptake [76]. Over prolonged drought, stomatal limitation may affect the balance between carbon supply from photosynthesis and consumption, leading to carbon starvation and consequent lack of energy to drive metabolism and repair damaged photosystems [76]. Thus, after prolonged drought, plants depleted in carbohydrates are more likely to incur into hydraulic failure, due to the inability in recovering damage, even after stress relief [77].

Recent evidences on poplar, highlighted that recovery of embolized conduits is a spatially coordinated and energy-demanding process, which requires the active translocation of sugars and other resources from the symplast to the apoplast of parenchyma cells [78]. These results confirm previous observations on grapevine [79] and support the studies speculating that different grapevines cultivars recover differently according to a divergent ability in translocating and utilizing nonstructural carbohydrates (NSCs) [20].

Since ABA is a key factor in modulating the starch-to-sugars pathway, by upregulating carbohydrate metabolism's enzymes (e.g., β-amylase and vacuolar invertase) [80,81], some studies have suggested that it may play a pivotal role in carbohydrate mobilization response during recovery [81–84]. Secchi et al. [82] applied exogenous ABA to ABA-deficient mutants of *Lycopersicon esculentum*, showing that increasing [ABA] favors embolized-vessels refilling. Differences in refilling were also correlated with petioles starch content, supporting the hypothesis of ABA being able to trigger sugars translocation for damage repairing. A detailed investigation on ABA role in sugars metabolism and mobilization has been recently conducted on *Populus nigra* by Brunetti et al. [83]. They observed that at recovery, bark-stocked starch is rapidly converted to soluble sugars, which are then translocated in the wood. Both starch depletion and soluble sugars increases have been correlated to [ABA], suggesting ABA being the trigger [83]. In grapevine, Perrone et al. [6] evaluated concurrently physiological response, ABA variation, and gene expression, in leaf petioles during water stress and subsequent recovery. Since the coordinated increment in [ABA] and expression of genes related to secondary metabolism during recovery, it is likely that ABA may play a pivotal role in coordinating the carbohydrates mobilization.

These studies increased our knowledge about the starch-to-NSCs conversion and subsequent mobilization during post-drought hydraulic recovery [85]. Furthermore, these observations may suggest a possible role of ABA in mediating processes related to carbohydrates metabolism where its role has not been considered so far. A recent study [20] pointed out that under drought, different grapevine cultivars display a contrasting pattern of NSCs utilization. The anisohydric Shiraz delayed NSCs consumption, and it implemented remarkable anatomical adjustments (decreased the size of xylem conduits). Conversely, in the near-isohydric Cabernet Sauvignon, stress response was based on an earlier starch depletion and NSCs mobilization [20].

The iso-/anisohydric classification has been redefined multiple times, the species/cultivar behaviors are distributed along a continuum rather than being dichotomous, and its variation depends on the plant interaction with the environment [60,65]. Nevertheless, the concept itself is useful to define the plant-specific stress management under long term [59]. Different grapevine cultivars have been sorted according to their iso-/anisohydric behavior; however, it has been observed that under different treatment/conditions, the same cultivar can display different iso/anisohydric behavior [51,60,86]. Since the increasing amount of evidences have linked ABA to a long-term regulation mechanism, and since it has been proved that differences in grapevine intraspecific anisohydricity have been linked to a variation in ABA biosynthesis [87], it seems likely that ABA may coordinate the ensemble of phenomena involved in stress response. Further investigation on the actual role of ABA is still needed, especially on poorly understood adaptation mechanisms as "stress memory" [88,89], which has also been related to modulations in carbohydrate metabolism [89–91] and have also been observed in grapevine [20,92].

5. ABA in Salt-Stress Response

Foreseen effects of climate change are expected both to increase the water consumption demand in agriculture and to decrease the abundance of water sources [2]. Therefore, the use of wastewaters for crops irrigation has been purposed [3]. Besides the reduction in freshwater use, the application of this strategy has led to many benefits as the improvement of nutrients recycling, the minimization of pollutants' discharge into waterways, and the increase in plant growth, photosynthesis activity, and carbohydrates production [3,9]. However, over long term, this strategy may be deleterious, because of the elevated salts content in wastewaters [93]. Excessive salts concentration in the circulating solution in soil alters the osmotic balance between roots and soil solution, imposing drought-like conditions for plants [94]. Furthermore, salts accumulation in plants tissues may reach toxic levels, hence having negative impacts on plant growth, leaf expansion, and photosynthetic efficiency [93].

Plants cope with saline conditions through an ensemble on biochemical strategies such as ions exclusion, compartmentalization (both at cellular and whole-plant level), synthesis of compatible solutes, change in photosynthetic pathway, and production of plant hormones as cytokinins and ABA [93]. In particular, ABA seems to play a pivotal role in response coordination under salt stress [95].

On the base of studies investigating salinity effects on plant varieties differing in ABA synthesis and sensitivity, it has been highlighted that plants have a set of genes regulating osmotic stress. The expression of these genes is triggered by ABA, which binds an ABA-response element (ABRE). Then, the accumulation of Ca^{2+} ions is induced in cytosol, which (coupled with reactive oxygen species under severe stress) act as secondary messengers triggering salt-response genes expression [9,95]. Effects related to ABA are many and include production of compatible osmolytes and antioxidants [9], Ca^{2+} uptake to maintain membranes' stability [96], reduced induction of leaf abscission by downregulating ethylene release and toxic Cl^- ions in leaves [97], change of membranes composition [9], and stomatal closure [95]. Moreover, high [ABA] during drought stress has been demonstrated to counteract the salt-induced photosynthesis downregulation [98], promote starch degradation, and coordinate carbohydrates mobilization [10].

In grapevine, salinity effects have been mainly investigated in arid or semiarid regions (e.g., Australia), where soils are naturally more saline, hence affecting vineyard cropping [99–101]. Despite, in the recent years, the interest in this topic has increased worldwide [102–104], studies investigating salt-stress responses in grapevine rather focus on plant performance in terms of fruit quality [105,106]. In fact, it has been demonstrated that the excess of salts causes accumulation of Cl^- and Na^+ ions in berry juice, negatively affecting fruit yield, berry quality, and wine production [99,100,107,108].

A recent study investigated the synergic effect of the partial root drying (PRD) irrigation techniques and moderate salinity conditions on Shiraz and Grenache grapevines, in order to evaluate whether PRD-induced stomatal closure may limit xylem loading of toxic ions [109]. Their results assessed that under PRD irrigation regime, vines were generally enriched in Na^+, K^+, Cl^-, and Ca^{2+} ions, if compared to well-watered (WW) plants with same salt conditions. However, in PRD vines Cl^- concentration was minor in leaves and higher in roots. These results combined with the lower roots biomass production under WW conditions, confirm previous observations claiming the root system being the pivotal site for ion-exclusion mechanisms [107–110]. Indeed, it has been demonstrated that salt uptake and translocation is regulated differently in different rootstocks, and the whole process depends on many mechanisms [111]. In 2010, Upreti and Murti [110] reported that under salt stress, more tolerant rootstocks accumulate more ABA. Despite these aspects have been poorly investigated, there are some evidences claiming that grapevine cultivars differing in anisohydric behavior differ in uptake and translocation of toxic ions [107,109,111,112].

Nowadays, literature focusing on ABA-mediated responses under salt stress is still scarce. However, the reported studies confirm the evidences on other species suggesting ABA being a pivotal hormone in salt-stress response mediation [113]. Hence, further investigations on this topic will be useful to better understand the underrated ABA role in this process.

6. Conclusions

ABA was widely investigated in the last decades due its role in regulating several physiological processes, including abiotic stress responses. While the understanding of biosynthetic pathway is consolidated, the triggering system of ABA signaling is still debated. In grapevine, thought hypothesis of ABA being mainly synthetized in the root system has been commonly accepted, a growing body of literature has been supporting the hypothesis that drought-induced stomatal closure is mediated by ABA produced into the leaf.

Another debated point is the ABA role at different intensities of stress: while there is consensus in the role of ABA in keeping stomata closed in order to inhibit premature stomata reopening after the end of water stress, its role in regulating the stomatal conductance at the onset of the stress is quite debated. Recent progress in the determination of ABA biosynthesis triggering mechanisms support the hypothesis of primary stomatal closure being induced by an ensemble of passive and active mechanisms unrelated to ABA, which seems to be mainly involved in more severe water stress response.

The dispute on the role of ABA in stomata regulation also involves the debate on the classification of grapevine genotypes in anisohydric and near-anisohydric categories. Investigations carried out in *V. vinifera* and other species have led to the conclusion that genotypes are not divided in two different categories but are distributed along a continuum, according to their degree of anisohydry. The physiological explanation of the different levels of anisohydry lies on the unraveling of the relative importance of the mechanisms inducing stomata regulation in response to the various environmental stimuli. Nevertheless, the attempt of phenotyping the genotypic response to drought is increasing in the last years, and it will provide useful information for viticulture as well as for the comprehension of grapevine physiology. An interesting topic in future research may involve the investigation of differences in the basal level of ABA and modulations of ABA content, in varieties differing for abiotic stress tolerance.

Although, in the last decades, the understanding of the physiological processes in which ABA is involved has notably increased, many aspects are still debated. In *V. vinifera*, future research efforts should be aimed at the comprehension of the role of ABA in drought stress in relation to water stress severity, duration, and frequency in order to make the experimental results more representative to what occurs in the field. Emerging topics deserving more attention are the interaction of ABA regulation and carbohydrates under water stress and the role of ABA in salt stress response, which is still poorly investigated, especially in this species.

Author Contributions: Original draft preparation, D.M.; writing, review and editing, D.M. and S.T. All authors have read and agreed to the published version of the manuscript.

Funding: This research received no external funding.

Conflicts of Interest: The authors declare no conflict of interest.

References

1. Choat, B.; Jansen, S.; Brodribb, T.J.; Cochard, H.; Delzon, S.; Bhaskar, R.; Bucci, S.J.; Feild, T.S.; Gleason, S.M.; Hacke, U.G.; et al. Global convergence in the vulnerability of forests to drought. *Nature* **2012**, *491*, 752–755. [CrossRef] [PubMed]
2. Stocker, T.F.; Qin, D.; Plattner, G.-K.; Tignor, M.; Allen, S.K.; Boschung, J.; Nauels, A.; Xia, Y.; Bex, V.; Midgley, P.M. *Climate Change 2013: The Physical Science Basis*; Cambridge University Press: Cambridge, UK, 2013.

3. Costa, J.M.; Vaz, M.; Escalona, J.; Egipto, R.; Lopes, C.; Medrano, H.; Chaves, M.M. Modern viticulture in southern Europe: Vulnerabilities and strategies for adaptation to water scarcity. *Agric. Water Manag.* **2016**, *164*, 5–18. [CrossRef]

4. Duchêne, E.; Huard, F.; Dumas, V.; Schneider, C.; Merdinoglu, D. The challenge of adapting grapevine varieties to climate change. *Clim. Res.* **2010**, *41*, 193–204. [CrossRef]

5. Gambetta, G.A.; Herrera, J.C.; Dayer, S.; Feng, Q.; Hochberg, U.; Castellarin, S.D. The physiology of drought stress in grapevine: Towards an integrative definition of drought tolerance. *J. Exp. Bot.* **2020**, *71*, 4658–4676. [CrossRef]

6. Perrone, I.; Pagliarani, C.; Lovisolo, C.; Chitarra, W.; Roman, F.; Schubert, A. Recovery from water stress affects grape leaf petiole transcriptome. *Planta* **2012**, *235*, 1383–1396. [CrossRef]

7. Savi, T.; Casolo, V.; Dal Borgo, A.; Rosner, S.; Torboli, V.; Stenni, B.; Bertoncin, P.; Martellos, S.; Pallavicini, A.; Nardini, A. Drought-induced dieback of *Pinus nigra*: A tale of hydraulic failure and carbon starvation. *Conserv. Physiol.* **2019**, *7*, coz012.

8. Lovisolo, C.; Perrone, I.; Carra, A.; Ferrandino, A.; Flexas, J.; Medrano, H.; Schubert, A. Drought-induced changes in development and function of grapevine (*Vitis* spp.) organs and in their hydraulic and non-hydraulic interactions at the whole-plant level: A physiological and molecular update. *Funct. Plant Biol.* **2010**, *37*, 98–116. [CrossRef]

9. Vishwakarma, K.; Upadhyay, N.; Kumar, N.; Yadav, G.; Singh, J.; Mishra, R.K.; Kumar, V.; Verma, R.; Upadhyay, R.G.; Pandey, M.; et al. Abscisic Acid Signaling and Abiotic Stress Tolerance in Plants: A Review on Current Knowledge and Future Prospects. *Front. Plant Sci.* **2017**, *8*, 161. [CrossRef]

10. Kempa, S.; Krasensky, J.; Santo, S.D.; Kopka, J.; Jonak, C. A Central Role of Abscisic Acid in Stress-Regulated Carbohydrate Metabolism. *PLoS ONE* **2008**, *3*, e3935. [CrossRef]

11. Millar, A.A.; Jacobsen, J.V.; Ross, J.J.; Helliwell, C.A.; Poole, A.T.; Scofield, G.; Reid, J.B.; Gubler, F. Seed dormancy and ABA metabolism in Arabidopsis and barley: The role of ABA 8′-hydroxylase. *Plant J.* **2006**, *45*, 942–954. [CrossRef]

12. Ding, Q.; Zeng, J.; He, X.-Q. MiR169 and its target PagHAP2-6 regulated by ABA are involved in poplar cambium dormancy. *J. Plant Physiol.* **2016**, *198*, 1–9. [CrossRef] [PubMed]

13. De Smet, I.; Signora, L.; Beeckman, T.; Inzé, D.; Foyer, C.H.; Zhang, H. An abscisic acid-sensitive checkpoint in lateral root development of *Arabidopsis*. *Plant J.* **2003**, *33*, 543–555. [CrossRef] [PubMed]

14. Jia, H.-F.; Chai, Y.-M.; Li, C.-L.; Lu, D.; Luo, J.-J.; Qin, L.; Shen, Y.-Y. Abscisic Acid Plays an Important Role in the Regulation of Strawberry Fruit Ripening. *Plant Physiol.* **2011**, *157*, 188–199. [CrossRef]

15. Downton, W.J.S.; Loveys, B.R.; Grant, W.J.R. Salinity effects on the stomatal behaviour of grapevine. *New Phytol.* **1990**, *116*, 499–503. [CrossRef]

16. Dodd, I.C.; Egea, G.; Davies, W.J. Abscisic acid signalling when soil moisture is heterogeneous: Decreased photoperiod sap flow from drying roots limits abscisic acid export to the shoots. *Plant Cell Environ.* **2008**, *31*, 1263–1274. [CrossRef]

17. Dodd, I.C.; Puértolas, J.; Huber, K.; Pérez-Pérez, J.G.; Wright, H.R.; Blackwell, M.S. The importance of soil drying and re-wetting in crop phytohormonal and nutritional responses to deficit irrigation. *J. Exp. Bot.* **2015**, *66*, 2239–2252. [CrossRef]

18. Frioni, T.; Tombesi, S.; Sabbatini, P.; Squeri, C.; Lavado Rodas, N.; Palliotti, A.; Poni, S. Kaolin Reduces ABA Biosynthesis through the Inhibition of Neoxanthin Synthesis in Grapevines under Water Deficit. *Int. J. Mol. Sci.* **2020**, *21*, 4950. [CrossRef]

19. Tombesi, S.; Nardini, A.; Frioni, T.; Soccolini, M.; Zadra, C.; Farinelli, D.; Poni, S.; Palliotti, A. Stomatal closure is induced by hydraulic signals and maintained by ABA in drought-stressed grapevine. *Sci. Rep.* **2015**, *5*, 12449. [CrossRef]

20. Falchi, R.; Petrussa, E.; Braidot, E.; Sivilotti, P.; Boscutti, F.; Vuerich, M.; Calligaro, C.; Filippi, A.; Herrera, J.C.; Sabbatini, P. Analysis of Non-Structural Carbohydrates and Xylem Anatomy of Leaf Petioles Offers New Insights in the Drought Response of Two Grapevine Cultivars. *Int. J. Mol. Sci.* **2020**, *21*, 1457. [CrossRef]

21. Saleh, B.; Alshehadah, E.; Slaman, H. Abscisic Acid (ABA) and Salicylic Acid (SA) Content in Relation to Transcriptional Patterns in Grapevine (*Vitis vinifera* L.) under Salt Stress. *J. Plant Biochem. Physiol.* **2020**, *8*, 245.

22. Banerjee, A.; Sharkey, T.D. Methylerythritol 4-phosphate (MEP) pathway metabolic regulation. *Nat. Prod. Rep.* **2014**, *31*, 1043–1055. [CrossRef] [PubMed]

23. Loveys, B.R.; Kriedemann, P.E. Internal Control of Stomatal Physiology and Photosynthesis. I. Stomatal Regulation and Associated Changes in Endogenous Levels of Abscisic and Phaseic Acids. *Funct. Plant Biol.* **1974**, *1*, 407–415. [CrossRef]

24. Loveys, B.R. The Intracellular Location of Abscisic Acid in Stressed and Non-Stressed Leaf Tissue. *Physiol. Plant.* **1977**, *40*, 6–10. [CrossRef]

25. Loveys, B.R.; Düring, H. Diurnal Changes in Water Relations and Abscisic Acid in Field-Grown *Vitis Vinifera* Cultivars. *New Phytol.* **1984**, *97*, 37–47. [CrossRef]

26. Pierce, M.; Raschke, K. Correlation between loss of turgor and accumulation of abscisic acid in detached leaves. *Planta* **1980**, *148*, 174–182. [CrossRef]

27. Zhang, J.; Davies, W.J. Increased Synthesis of ABA in Partially Dehydrated Root Tips and ABA Transport from Roots to Leaves. *J. Exp. Bot.* **1987**, *38*, 2015–2023. [CrossRef]

28. Zhang, J.; Davies, W.J. Changes in the concentration of ABA in xylem sap as a function of changing soil water status can account for changes in leaf conductance and growth. *Plant Cell Environ.* **1990**, *13*, 277–285. [CrossRef]

29. Soar, C.J.; Dry, P.R.; Loveys, B.R. Scion photosynthesis and leaf gas exchange in *Vitis vinifera* L. cv. Shiraz: Mediation of rootstock effects via xylem sap ABA. *Aust. J. Grape Wine Res.* **2006**, *12*, 82–96. [CrossRef]

30. Speirs, J.; Binney, A.; Collins, M.; Edwards, E.; Loveys, B. Expression of ABA synthesis and metabolism genes under different irrigation strategies and atmospheric VPDs is associated with stomatal conductance in grapevine (*Vitis vinifera* L. cv Cabernet Sauvignon). *J. Exp. Bot.* **2013**, *64*, 1907–1916. [CrossRef]

31. Dry, P.R.; Loveys, B.R.; During, H. Partial drying of the rootzone of grape I. Transient changes in shoot growth and gas exchange. *Vitis-Geilweilerhof-* **2000**, *39*, 3–8.

32. Poni, S.; Bernizzoni, F.; Civardi, S. Response of "Sangiovese" grapevines to partial root-zone drying: Gas-exchange, growth and grape composition. *Sci. Hortic.* **2007**, *114*, 96–103. [CrossRef]

33. Romero, P.; Dodd, I.C.; Martinez-Cutillas, A. Contrasting physiological effects of partial root zone drying in field-grown grapevine (*Vitis vinifera* L. cv. Monastrell) according to total soil water availability. *J. Exp. Bot.* **2012**, *63*, 4071–4083. [CrossRef] [PubMed]

34. Romero, P.; Pérez-Pérez, J.G.; Amor, F.M.D.; Martinez-Cutillas, A.; Dodd, I.C.; Botía, P. Partial root zone drying exerts different physiological responses on field-grown grapevine (*Vitis vinifera* cv. Monastrell) in comparison to regulated deficit irrigation. *Funct. Plant Biol.* **2014**, *41*, 1087–1106. [CrossRef] [PubMed]

35. Schachtman, D.P.; Goodger, J.Q.D. Chemical root to shoot signaling under drought. *Trends Plant Sci.* **2008**, *13*, 281–287. [CrossRef]

36. Dodd, I.C.; Theobald, J.C.; Bacon, M.A.; Davies, W.J. Alternation of wet and dry sides during partial rootzone drying irrigation alters root-to-shoot signalling of abscisic acid. *Funct. Plant Biol.* **2006**, *33*, 1081–1089. [CrossRef]

37. Kochhar, S.L.; Gujral, S.K. *Plant Physiology: Theory and Applications: Theory and Applications*; Cambridge University Press: Cambridge, UK, 2020; pp. 491–495.

38. Pérez-Pérez, J.G.; Puertolas, J.; Albacete, A.; Dodd, I.C. Alternation of wet and dry sides during partial rootzone drying irrigation enhances leaf ethylene evolution. *Environ. Exp. Bot.* **2020**, *176*, 104095. [CrossRef]

39. Soar, C.J.; Speirs, J.; Maffei, S.M.; Penrose, A.B.; Mccarthy, M.G.; Loveys, B.R. Grape vine varieties Shiraz and Grenache differ in their stomatal response to VPD: Apparent links with ABA physiology and gene expression in leaf tissue. *Aust. J. Grape Wine Res.* **2006**, *12*, 2–12. [CrossRef]

40. Ikegami, K.; Okamoto, M.; Seo, M.; Koshiba, T. Activation of abscisic acid biosynthesis in the leaves of *Arabidopsis thaliana* in response to water deficit. *J. Plant Res.* **2008**, *122*, 235. [CrossRef]

41. Manzi, M.; Lado, J.; Rodrigo, M.J.; Zacarías, L.; Arbona, V.; Gómez-Cadenas, A. Root ABA Accumulation in Long-Term Water-Stressed Plants is Sustained by Hormone Transport from Aerial Organs. *Plant Cell Physiol.* **2015**, *56*, 2457–2466. [CrossRef]

42. McAdam, S.A.M.; Brodribb, T.J. Mesophyll Cells Are the Main Site of Abscisic Acid Biosynthesis in Water-Stressed Leaves. *Plant Physiol.* **2018**, *177*, 911–917. [CrossRef]

43. Sack, L.; John, G.P.; Buckley, T.N. ABA Accumulation in Dehydrating Leaves Is Associated with Decline in Cell Volume, Not Turgor Pressure. *Plant Physiol.* **2018**, *176*, 489–495. [CrossRef] [PubMed]

44. Castro, P.; Puertolas, J.; Dodd, I.C. Stem girdling uncouples soybean stomatal conductance from leaf water potential by enhancing leaf xylem ABA concentration. *Environ. Exp. Bot.* **2019**, *159*, 149–156. [CrossRef]

45. Li, L.; Yuan, H. Chromoplast biogenesis and carotenoid accumulation. *Arch. Biochem. Biophys.* **2013**, *539*, 102–109. [CrossRef] [PubMed]

46. Howitt, C.A.; Pogson, B.J. Carotenoid accumulation and function in seeds and non-green tissues. *Plant Cell Environ.* **2006**, *29*, 435–445. [CrossRef] [PubMed]

47. Kuromori, T.; Fujita, M.; Urano, K.; Tanabata, T.; Sugimoto, E.; Shinozaki, K. Overexpression of AtABCG25 enhances the abscisic acid signal in guard cells and improves plant water use efficiency. *Plant Sci.* **2016**, *251*, 75–81. [CrossRef] [PubMed]

48. Bota, B.J.; Flexas, J.; Medrano, H. Genetic variability of photosynthesis and water use in Balearic grapevine cultivars. *Ann. Appl. Biol.* **2001**, *138*, 353–361. [CrossRef]

49. Levin, A.D.; Williams, L.E.; Matthews, M.A. A continuum of stomatal responses to water deficits among 17 wine grape cultivars (*Vitis vinifera*). *Funct. Plant Biol.* **2020**, *47*, 11–25. [CrossRef]

50. Pagliarani, C.; Vitali, M.; Ferrero, M.; Vitulo, N.; Incarbone, M.; Lovisolo, C.; Valle, G.; Schubert, A. The Accumulation of miRNAs Differentially Modulated by Drought Stress Is Affected by Grafting in Grapevine. *Plant Physiol.* **2017**, *173*, 2180–2195. [CrossRef]

51. Bota, J.; Tomás, M.; Flexas, J.; Medrano, H.; Escalona, J.M. Differences among grapevine cultivars in their stomatal behavior and water use efficiency under progressive water stress. *Agric. Water Manag.* **2016**, *164*, 91–99. [CrossRef]

52. Tyree, M.T. The Cohesion-Tension theory of sap ascent: Current controversies. *J. Exp. Bot.* **1997**, *48*, 1753–1765. [CrossRef]

53. Tyree, M.T.; Sperry, J.S. Vulnerability of Xylem to Cavitation and Embolism. *Annu. Rev. Plant Physiol. Plant Mol. Biol.* **1989**, *40*, 19–36. [CrossRef]

54. Hacke, U.G.; Sperry, J.S.; Pockman, W.T.; Davis, S.D.; McCulloh, K.A. Trends in wood density and structure are linked to prevention of xylem implosion by negative pressure. *Oecologia* **2001**, *126*, 457–461. [CrossRef] [PubMed]

55. Buckley, T.N. How do stomata respond to water status? *New Phytol.* **2019**, *224*, 21–36. [CrossRef] [PubMed]

56. Mäser, P.; Leonhardt, N.; Schroeder, J. The Clickable Guard Cell: Electronically Linked Model of Guard Cell Signal Transduction Pathways. *Arab. Book* **2003**, *32*, 1–4.

57. Wang, Y.; Chen, Z.-H.; Zhang, B.; Hills, A.; Blatt, M.R. PYR/PYL/RCAR Abscisic Acid Receptors Regulate K$^+$ and Cl$^-$ Channels through Reactive Oxygen Species-Mediated Activation of Ca^{2+} Channels at the Plasma Membrane of Intact Arabidopsis Guard Cells. *Plant Physiol.* **2013**, *163*, 566–577. [CrossRef]

58. Tardieu, F.; Simonneau, T. Variability among species of stomatal control under fluctuating soil water status and evaporative demand: Modelling isohydric and anisohydric behaviours. *J. Exp. Bot.* **1998**, *49*, 419–432. [CrossRef]

59. Ratzmann, G.; Meinzer, F.C.; Tietjen, B. Iso/Anisohydry: Still a Useful Concept. *Trends Plant Sci.* **2019**, *24*, 191–194. [CrossRef]

60. Hochberg, U.; Rockwell, F.E.; Holbrook, N.M.; Cochard, H. Iso/Anisohydry: A Plant–Environment Interaction Rather Than a Simple Hydraulic Trait. *Trends Plant Sci.* **2018**, *23*, 112–120. [CrossRef]

61. Coupel-Ledru, A.; Lebon, É.; Christophe, A.; Doligez, A.; Cabrera-Bosquet, L.; Péchier, P.; Hamard, P.; This, P.; Simonneau, T. Genetic variation in a grapevine progeny (*Vitis vinifera* L. cvs Grenache × Syrah) reveals inconsistencies between maintenance of daytime leaf water potential and response of transpiration rate under drought. *J. Exp. Bot.* **2014**, *65*, 6205–6218. [CrossRef]

62. Nardini, A.; Salleo, S. Limitation of stomatal conductance by hydraulic traits: Sensing or preventing xylem cavitation? *Trees* **2000**, *15*, 14–24. [CrossRef]

63. Christmann, A.; Weiler, E.W.; Steudle, E.; Grill, E. A hydraulic signal in root-to-shoot signalling of water shortage. *Plant J.* **2007**, *52*, 167–174. [CrossRef] [PubMed]

64. Hubbard, R.M.; Bond, B.J.; Ryan, M.G. Evidence that hydraulic conductance limits photosynthesis in old *Pinus ponderosa* trees. *Tree Physiol.* **1999**, *19*, 165–172. [CrossRef] [PubMed]

65. Klein, T. The variability of stomatal sensitivity to leaf water potential across tree species indicates a continuum between isohydric and anisohydric behaviours. *Funct. Ecol.* **2014**, *28*, 1313–1320. [CrossRef]

66. Tombesi, S.; Nardini, A.; Farinelli, D.; Palliotti, A. Relationships between stomatal behavior, xylem vulnerability to cavitation and leaf water relations in two cultivars of *Vitis vinifera*. *Physiol. Plant.* **2014**, *152*, 453–464. [CrossRef] [PubMed]

67. Skelton, R.P.; West, A.G.; Dawson, T.E. Predicting plant vulnerability to drought in biodiverse regions using functional traits. *Proc. Natl. Acad. Sci. USA* **2015**, *112*, 5744–5749. [CrossRef] [PubMed]

68. Hochberg, U.; Windt, C.W.; Ponomarenko, A.; Zhang, Y.-J.; Gersony, J.; Rockwell, F.E.; Holbrook, N.M. Stomatal Closure, Basal Leaf Embolism, and Shedding Protect the Hydraulic Integrity of Grape Stems. *Plant Physiol.* **2017**, *174*, 764–775. [CrossRef]

69. McAdam, S.A.M.; Brodribb, T.J. Separating Active and Passive Influences on Stomatal Control of Transpiration. *Plant Physiol.* **2014**, *164*, 1578–1586. [CrossRef]

70. Degu, A.; Hochberg, U.; Wong, D.C.; Alberti, G.; Lazarovitch, N.; Peterlunger, E.; Castellarin, S.D.; Herrera, J.C.; Fait, A. Swift metabolite changes and leaf shedding are milestones in the acclimation process of grapevine under prolonged water stress. *BMC Plant Biol.* **2019**, *19*, 69. [CrossRef]

71. Tramontini, S.; Döring, J.; Vitali, M.; Ferrandino, A.; Stoll, M.; Lovisolo, C. Soil water-holding capacity mediates hydraulic and hormonal signals of near-isohydric and near-anisohydric *Vitis* cultivars in potted grapevines. *Funct. Plant Biol.* **2014**, *41*, 1119–1128. [CrossRef]

72. Coupel-Ledru, A.; Tyerman, S.D.; Masclef, D.; Lebon, E.; Christophe, A.; Edwards, E.J.; Simonneau, T. Abscisic acid down-regulates hydraulic conductance of grapevine leaves in isohydric genotypes only. *Plant Physiol.* **2017**, *175*, 1121–1134. [CrossRef]

73. Dayer, S.; Scharwies, J.D.; Ramesh, S.A.; Sullivan, W.; Doerflinger, F.C.; Pagay, V.; Tyerman, S.D. Comparing hydraulics between two grapevine cultivars reveals differences in stomatal regulation under water stress and exogenous ABA applications. *Front. Plant Sci.* **2020**, *11*, 705. [CrossRef] [PubMed]

74. Gambetta, G.A.; Knipfer, T.; Fricke, W.; McElrone, A.J. Aquaporins and Root Water Uptake. In *Plant Aquaporins: From Transport to Signaling*; Chaumont, F., Tyerman, S.D., Eds.; Signaling and Communication in Plants; Springer International Publishing: Cham, Switzerland, 2017; pp. 133–153.

75. Dayer, S.; Reingwirtz, I.; McElrone, A.J.; Gambetta, G.A. Response and Recovery of Grapevine to Water Deficit: From Genes to Physiology. In *The Grape Genome*; Springer: Berlin/Heidelberg, Germany, 2019; pp. 223–245.

76. McDowell, N.G.; Beerling, D.J.; Breshears, D.D.; Fisher, R.A.; Raffa, K.F.; Stitt, M. The interdependence of mechanisms underlying climate-driven vegetation mortality. *Trends Ecol. Evol.* **2011**, *26*, 523–532. [CrossRef] [PubMed]

77. Tomasella, M.; Petrussa, E.; Petruzzellis, F.; Nardini, A.; Casolo, V. The Possible Role of Non-Structural Carbohydrates in the Regulation of Tree Hydraulics. *Int. J. Mol. Sci.* **2020**, *21*, 144. [CrossRef] [PubMed]

78. Secchi, F.; Pagliarani, C.; Cavalletto, S.; Petruzzellis, F.; Tonel, G.; Savi, T.; Tromba, G.; Obertino, M.M.; Lovisolo, C.; Nardini, A.; et al. Chemical inhibition of xylem cellular activity impedes the removal of drought-induced embolisms in poplar stems—New insights from micro-CT analysis. *New Phytol.* **2020**. [CrossRef]

79. Brodersen, C.R.; McElrone, A.J.; Choat, B.; Lee, E.F.; Shackel, K.A.; Matthews, M.A. In vivo visualizations of drought-induced embolism spread in *Vitis vinifera*. *Plant Physiol.* **2013**, *161*, 1820–1829. [CrossRef]

80. Pan, Q.-H.; Li, M.-J.; Peng, C.-C.; Zhang, N.; Zou, X.; Zou, K.-Q.; Wang, X.-L.; Yu, X.-C.; Wang, X.-F.; Zhang, D.-P. Abscisic acid activates acid invertases in developing grape berry. *Physiol. Plant.* **2005**, *125*, 157–170. [CrossRef]

81. Thalmann, M.; Pazmino, D.; Seung, D.; Horrer, D.; Nigro, A.; Meier, T.; Kölling, K.; Pfeifhofer, H.W.; Zeeman, S.C.; Santelia, D. Regulation of Leaf Starch Degradation by Abscisic Acid Is Important for Osmotic Stress Tolerance in Plants. *Plant Cell* **2016**, *28*, 1860–1878. [CrossRef]

82. Secchi, F.; Perrone, I.; Chitarra, W.; Zwieniecka, A.K.; Lovisolo, C.; Zwieniecki, M.A. The Dynamics of Embolism Refilling in Abscisic Acid (ABA)-Deficient Tomato Plants. *Int. J. Mol. Sci.* **2013**, *14*, 359–377. [CrossRef]

83. Brunetti, C.; Savi, T.; Nardini, A.; Loreto, F.; Gori, A.; Centritto, M. Changes in abscisic acid content during and after drought are related to carbohydrate mobilization and hydraulic recovery in poplar stems. *Tree Physiol.* **2020**, *40*, 1043–1057. [CrossRef]

84. Cardoso, A.A.; Gori, A.; Da-Silva, C.J.; Brunetti, C. Abscisic Acid Biosynthesis and Signaling in Plants: Key Targets to Improve Water Use Efficiency and Drought Tolerance. *Appl. Sci.* **2020**, *10*, 6322. [CrossRef]

85. Tomasella, M.; Häberle, K.-H.; Nardini, A.; Hesse, B.; Machlet, A.; Matyssek, R. Post-drought hydraulic recovery is accompanied by non-structural carbohydrate depletion in the stem wood of Norway spruce saplings. *Sci. Rep.* **2017**, *7*, 14308. [CrossRef]

86. Tortosa, I.; Escalona, J.M.; Douthe, C.; Pou, A.; Garcia-Escudero, E.; Toro, G.; Medrano, H. The intra-cultivar variability on water use efficiency at different water status as a target selection in grapevine: Influence of ambient and genotype. *Agric. Water Manag.* **2019**, *223*, 105648. [CrossRef]

87. Rogiers, S.Y.; Hardie, W.J.; Smith, J.P. Stomatal density of grapevine leaves (*Vitis vinifera* L.) responds to soil temperature and atmospheric carbon dioxide. *Aust. J. Grape Wine Res.* **2011**, *17*, 147–152. [CrossRef]

88. Walter, J.; Nagy, L.; Hein, R.; Rascher, U.; Beierkuhnlein, C.; Willner, E.; Jentsch, A. Do plants remember drought? Hints towards a drought-memory in grasses. *Environ. Exp. Bot.* **2011**, *71*, 34–40. [CrossRef]

89. Lukić, N.; Kukavica, B.; Davidović-Plavšić, B.; Hasanagić, D.; Walter, J. Plant stress memory is linked to high levels of anti-oxidative enzymes over several weeks. *Environ. Exp. Bot.* **2020**, *178*, 104166. [CrossRef]

90. Li, Y.; Xu, S.; Wang, Z.; He, L.; Xu, K.; Wang, G. Glucose triggers stomatal closure mediated by basal signaling through HXK1 and PYR/RCAR receptors in Arabidopsis. *J. Exp. Bot.* **2018**, *69*, 1471–1484. [CrossRef]

91. Nguyen, T.T.Q.; Trinh, L.T.H.; Pham, H.B.V.; Le, T.V.; Phung, T.K.H.; Lee, S.-H.; Cheong, J.-J. Evaluation of proline, soluble sugar and ABA content in soybean *Glycine max* (L.) under drought stress memory. *AIMS Bioeng.* **2020**, *7*, 114. [CrossRef]

92. Netzer, Y.; Munitz, S.; Shtein, I.; Schwartz, A. Structural memory in grapevines: Early season water availability affects late season drought stress severity. *Eur. J. Agron.* **2019**, *105*, 96–103. [CrossRef]

93. Parida, A.K.; Das, A.B. Salt tolerance and salinity effects on plants: A review. *Ecotoxicol. Environ. Saf.* **2005**, *60*, 324–349. [CrossRef]

94. Laurenson, S.; Bolan, N.S.; Smith, E.; Mccarthy, M. Review: Use of recycled wastewater for irrigating grapevines. *Aust. J. Grape Wine Res.* **2012**, *18*, 1–10. [CrossRef]

95. Zhang, J.; Jia, W.; Yang, J.; Ismail, A.M. Role of ABA in integrating plant responses to drought and salt stresses. *Field Crop. Res.* **2006**, *97*, 111–119. [CrossRef]

96. Chen, S.; Li, J.; Wang, S.; Hüttermann, A.; Altman, A. Salt, nutrient uptake and transport, and ABA of *Populus euphratica*; a hybrid in response to increasing soil NaCl. *Trees* **2001**, *15*, 186–194. [CrossRef]

97. Gomez-Cadenas, A.; Arbona, V.; Jacas, J.; Primo-Millo, E.; Talon, M. Abscisic acid reduces leaf abscission and increases salt tolerance in *citrus* plants. *J. Plant Growth Regul.* **2002**, *21*, 234–240. [CrossRef]

98. Popova, L.P.; Stoinova, Z.G.; Maslenkova, L.T. Involvement of abscisic acid in photosynthetic process in *Hordeum vulgare* L. during salinity stress. *J. Plant Growth Regul.* **1995**, *14*, 211. [CrossRef]

99. Downton, W.J.S. Influence of rootstocks on the accumulation of chloride, sodium and potassium in grapevines. *Aust. J. Agric. Res.* **1977**, *28*, 879–889. [CrossRef]

100. Stevens, R.M.; Harvey, G.; Partington, D.L. Irrigation of grapevines with saline water at different growth stages: Effects on leaf, wood and juice composition. *Aust. J. Grape Wine Res.* **2011**, *17*, 239–248. [CrossRef]

101. Walker, R.R.; Blackmore, D.H.; Clingeleffer, P.R.; Emanuelli, D. Rootstock type determines tolerance of Chardonnay and Shiraz to long-term saline irrigation. *Aust. J. Grape Wine Res.* **2014**, *20*, 496–506. [CrossRef]

102. Hassena, A.B.; Zouari, M.; Trabelsi, L.; Khabou, W.; Zouari, N. Physiological improvements of young olive tree (*Olea europaea* L. cv. Chetoui) under short term irrigation with treated wastewater. *Agric. Water Manag.* **2018**, *207*, 53–58. [CrossRef]

103. Martin, L.; Vila, H.; Bottini, R.; Berli, F. Rootstocks increase grapevine tolerance to NaCl through ion compartmentalization and exclusion. *Acta Physiol. Plant.* **2020**, *42*, 145. [CrossRef]

104. Aydemir, B.Ç.; Özmen, C.Y.; Kibar, U.; Mutaf, F.; Büyük, P.B.; Bakır, M.; Ergül, A. Salt stress induces endoplasmic reticulum stress-responsive genes in a grapevine rootstock. *PLoS ONE* **2020**, *15*, e0236424.

105. Li, F.; Zhao, X.; Yang, X.; He, X. Influence of NaCl stress on ABA synthesis and hardness of *Vitis vinifera* L.'Cabernet Sauvignon' berry in veraison. In Proceedings of the AIP Conference Proceedings, Proceedings of the 2nd International Conference on Frontiers of Biological Sciences and Engineering (FBSE 2019), Jinan, China, 19–20 October 2019; AIP Publishing LLC: Melville, NY, USA, 2020; Volume 2208, p. 020002.

106. Zhao, P.; Yang, X.; Han, N.; He, X. Influence of salt-alkali stress on quality formation of *Vitis vinifera* L.'Cabernet Sauvignon' of wine grape. In Proceedings of the AIP Conference Proceedings, Proceedings of the 2nd International Conference on Frontiers of Biological Sciences and Engineering (FBSE 2019), Jinan, China, 19–20 October 2019; AIP Publishing LLC: Melville, NY, USA, 2020; Volume 2208, p. 020006.

107. Walker, R.R.; Read, P.E.; Blackmore, D.H. Rootstock and salinity effects on rates of berry maturation, ion accumulation and colour development in Shiraz grapes. *Aust. J. Grape Wine Res.* **2000**, *6*, 227–239. [CrossRef]

108. Teakle, N.L.; Tyerman, S.D. Mechanisms of Cl⁻ transport contributing to salt tolerance. *Plant Cell Environ.* **2010**, *33*, 566–589. [CrossRef]

109. Degaris, K.A.; Walker, R.R.; Loveys, B.R.; Tyerman, S.D. Comparative effects of deficit and partial root-zone drying irrigation techniques using moderately saline water on ion partitioning in Shiraz and Grenache grapevines. *Aust. J. Grape Wine Res.* **2016**, *22*, 296–306. [CrossRef]
110. Upreti, K.K.; Murti, G.S.R. Response of grape rootstocks to salinity: Changes in root growth, polyamines and abscisic acid. *Biol. Plant.* **2010**, *54*, 730–734. [CrossRef]
111. Paranychianakis, N.V.; Angelakis, A.N. The effect of water stress and rootstock on the development of leaf injuries in grapevines irrigated with saline effluent. *Agric. Water Manag.* **2008**, *95*, 375–382. [CrossRef]
112. Vincent, D.; Ergül, A.; Bohlman, M.C.; Tattersall, E.A.R.; Tillett, R.L.; Wheatley, M.D.; Woolsey, R.; Quilici, D.R.; Joets, J.; Schlauch, K.; et al. Proteomic analysis reveals differences between *Vitis vinifera* L. cv. Chardonnay and cv. Cabernet Sauvignon and their responses to water deficit and salinity. *J. Exp. Bot.* **2007**, *58*, 1873–1892. [CrossRef]
113. Pye, M.F.; Dye, S.M.; Resende, R.S.; MacDonald, J.D.; Bostock, R.M. Abscisic acid as a dominant signal in tomato during salt stress predisposition to phytophthora root and crown rot. *Front. Plant Sci.* **2018**, *9*, 525. [CrossRef]

Publisher's Note: MDPI stays neutral with regard to jurisdictional claims in published maps and institutional affiliations.

International Journal of
Molecular Sciences

Review

Abscisic Acid and Flowering Regulation: Many Targets, Different Places

Damiano Martignago , **Beata Siemiatkowska** , **Alessandra Lombardi and Lucio Conti** *

Department of Biosciences, University of Milan, Via Giovanni Celoria, 26-20133 Milan, Italy;
damiano.martignago@unimi.it (D.M.); beata.siemiatkowska@unimi.it (B.S.); alessandra.lombardi@unimi.it (A.L.)
* Correspondence: lucio.conti@unimi.it; Tel.: +39-02-5031-5012

Received: 24 November 2020; Accepted: 17 December 2020; Published: 18 December 2020

Abstract: Plants can react to drought stress by anticipating flowering, an adaptive strategy for plant survival in dry climates known as drought escape (DE). In Arabidopsis, the study of DE brought to surface the involvement of abscisic acid (ABA) in controlling the floral transition. A central question concerns how and in what spatial context can ABA signals affect the floral network. In the leaf, ABA signaling affects flowering genes responsible for the production of the main florigen FLOWERING LOCUS T (FT). At the shoot apex, FD and FD-like transcription factors interact with FT and FT-like proteins to regulate ABA responses. This knowledge will help separate general and specific roles of ABA signaling with potential benefits to both biology and agriculture.

Keywords: abscisic acid (ABA); flowering time; Arabidopsis; drought escape; drought; *bZIP*; *GIGANTEA*; *CONSTANS*; *FLOWERING LOCUS T*; *FD*

1. Introduction

Plant hormone signaling pathways are highly interconnected to allow plants to finely adjust growth and development according to varying environmental stimuli derived from growth conditions, nutrient availability, biotic and abiotic stress [1]. Abscisic acid (ABA) is long known to play central roles in drought, osmotic, and high salinity responses, hence generally considered as a stress-related hormone [2]. One of the best-characterized mechanism of action of ABA in response to drought stress is the control of transpiration via stomatal opening and closure [3]. However, there is a growing body of evidence that points to ABA involvement in plant growth and developmental processes well beyond stress responses. In well-watered, nonstressed conditions, ABA signaling is required in root tissues for growth, hydrotropism, xylem formation, and suberin deposition, in the leaves for leaf initiation and development (as reviewed in [4]). In this review, we will explore known and potential modes of interaction between ABA signaling and the genes that control the transition to flowering. To support the reader, we will introduce some specific notions related to the regulation of the floral transition, ABA biosynthesis and signaling while referring to more specialized readings whenever it will be required.

Day Length Is a Key Floral Trigger in Arabidopsis

The transition to flowering marks the switch from the vegetative to the reproductive stage. Most plant species need to commit to flowering in a short window of time during the year to ensure optimal reproductive success. Hence, the timing of this transition is highly sensitive to environmental factors, enabling plants to align pollination, fruit and seed development with the most favorable conditions. Seasonal variations in mean temperature, day length (also known as photoperiod), and availability of nutrients and water are among the known environmental cues that regulate flowering.

Four floral pathways have been primarily described in the model plant *Arabidopsis thaliana* (Arabidopsis) through different genetic screens coupled with physiological analyses [5]. These pathways convey signals from photoperiod, vernalization, different endogenous cues as well as gibberellic acid (GA) accumulation, which is also linked to the age pathway [6]. Because of the established connections with ABA signaling, we will first focus on the photoperiodic pathway.

Most plants are sensitive to variations in photoperiod, which act as a critical seasonal cue at temperate latitudes [7]. In Arabidopsis, long day conditions promote flowering via activation of a signaling cascade that converges to the transcription of *FLOWERING LOCUS T* (*FT*) and *TWIN SISTER OF FT* (*TSF*) in the phloem companion cells [8]. Upon export from the phloem companion cells to sieve elements [9], FT and TSF gene products act as a florigenic signal, moving via the phloematic stream towards the shoot apex [10–12]. The photoperiodic cascade is activated upon exposure of plants to long-day conditions when GIGANTEA (GI) and the blue light receptor FLAVIN-BINDING, KELCH REPEAT, F-BOX1 (FKF1) display similar diel-regulated accumulations in the light phase [13,14]. The light-stabilized GI-FKF1 complex triggers the proteasomal degradation of CYCLING DOF FACTORs (CDFs), a family of transcriptional regulators that repress *CONSTANS* (*CO*) [15]. The diel degradation of CDFs alongside several mechanisms that regulate *CO* transcript accumulation contribute to a robust daytime expression of *CO* message, peaking at dusk [16]. Light is necessary to stabilize the CO protein, and numerous light-dependent molecular mechanisms involved in this process have been described [13]. CO protein plays a key role in the transcriptional activation of the florigens, forming a trimeric complex with NUCLEAR FACTOR-Y (NF-Y) B and C subunits at *CONSTANS RESPONSIVE* elements (CORE) located at the promoter of *FT* [17,18]. NF-Ys are highly conserved trimeric transcription factors (TFs) formed by NF-YA, NF-YB, and NF-YC subunits involved in many developmental processes including flowering [19]. Furthermore, *FT* transcript levels are further regulated by many transcriptional events that respond to a vast array of environmental and endogenous signals [20–22]. Thus, while typically CO is required for *FT* transcriptional activation, different TFs enable the fine-tuning of *FT* levels appropriate to the environmental conditions, thereby conferring substantial plasticity to the flowering process.

The mobilization of FT at the shoot apex triggers a change in the shoot identity that switches from producing leaves to floral primordia. As described in the current model for FT signaling, upon its relocation at the apex, FT protein forms a complex with the basic leucine zipper (bZIP) domain TF FD, probably with the participation of 14-3-3 proteins [23,24]. Although FD can bind DNA on its own, heterodimerization with FT (or TSF) and its phosphorylation enhance its DNA binding activity [24]. The function of FT is antagonized by a structurally related protein, TERMINAL FLOWER 1 (TFL1), which is also mobile, albeit its range of movement appears to be limited to the shoot meristem cells [25,26]. TFL1 can interact with FD and is recruited via FD at thousands of genomic positions where it exerts transcriptional repression [27,28]. FT outcompetes TFL1 for FD binding and thus activates transcription of FD targets which include floral meristem identity genes, conferring a floral fate to newly arising lateral primordia, and hormone-related gene functions [27,29].

In addition to the photoperiodic pathway, winter-annual accessions of Arabidopsis require the experience of cold to flower in the following spring, a process referred to as vernalization. This annual habit is conferred by two loci, the floral repressor *FLOWERING LOCUS C* (*FLC*), a MADS-box type TF and its upstream activator *FRIGIDA* (*FRI*) encoding a coiled-coil domain protein acting as transcriptional regulator and chromatin modifier [30]. Exposure of vernalization-sensitive *FRI FLC* seedlings to cold temperature triggers the epigenetic silencing of *FLC*, which is mediated by several chromatin remodeling proteins [31]. The repressed state of *FLC* chromatin causes its transcriptional inactivation and is maintained through mitotic cell divisions upon a return to warm temperature. Misexpression studies allowed to define the spatial interactions between the photoperiod pathway and vernalization response. *FLC* represses *FT* in the leaf and several floral genes expressed at the shoot meristem including *FD* [32]. Thus, vernalization enables the transcriptional activation and mobilization of the main systemic flowering signal FT and its response in shoot meristem cells.

Flowering in Arabidopsis is also positively regulated by gibberellins which play an essential role under noninductive short-day conditions [33–35]. GA signaling is mediated by a class of proteins named DELLA that act as negative regulators of GA responses. DELLAs interact with a vast array of proteins (mainly TFs) which preside different hormonal and developmental processes [36]. According to a consolidated model, DELLA binding usually impairs TFs function or their DNA accessibility which blocks GA-regulated transcriptional events [37]. An increase in GAs cellular concentration triggers a signaling cascade that leads to DELLA ubiquitination and its proteasomal degradation, thus promoting TFs function. For these reasons, the GA pathway, via control of DELLA levels, plays a key integrative role by modulating multiple floral inputs in different spatial contexts [38].

2. ABA Signaling and Its Multiple Connections with the Photoperiodic Pathway

Water deficit conditions experienced by Arabidopsis during the vegetative phase result in accelerated flowering compared to normal watering conditions [39]. This plastic shift in flowering activation is considered adaptive and referred to as drought escape (DE), a bet-hedging strategy that enables plants to attain reproductive development and achieve an early seed set under water-scarce environments [40]. While succeeding in reproduction under potentially lethal drought conditions, the cost associated with this strategy is a considerable reduction in seed number production as a result of shortened vegetative growth [41].

Genetic screens identified several mutants impaired in DE, the vast majority of which are defective in the photoperiodic response [42,43]. Consistent with the requirement of long-day-stimulated photoperiodic signaling in DE activation, water deficit conditions applied under short days do not cause DE (conversely, they delay flowering). Accordingly, increased levels of florigen *FT* and *TSF* accumulate in response to water deficit only under long-day photoperiods. Thus, drought signals can be interpreted as positive cues for flowering depending on the activation status of the photoperiodic cascade. GI is required in this process, as no florigen expression occurs in *gi* mutants under any photoperiodic regime [39]. This initial model has been further refined to indicate that GI is not just indirectly required to activate the photoperiodic cascade (e.g., through the transcriptional activation of *CO*) [43]. Indeed, GI conveys drought-derived cues upstream of *FT* in parallel to CO (this aspect will be discussed in more detail below).

As the basic structure of the DE process is highly intertwined with the photoperiodic genes, the integration of drought stimuli with the floral network is in large part mediated by ABA. In flowering plants, ABA is synthesized via the carotenoid pathway by cleavage of β-carotene metabolites called xanthophylls and shares the same intermediate molecular pool of other plant hormones like cytokinins, brassinosteroids, and GA. The first steps of ABA biosynthesis take place in the plastid, with the oxidative cleavage of zeaxanthin into all-*trans*-violaxanthin by the enzyme zeaxanthin epoxidase, encoded in Arabidopsis by ABA DEFICIENT 1 (ABA1). The *NINE-CIS-EPOXYCAROTENOID DIOXYGENASES* (NCEDs) produce the C_{15} xanthoxin which is translocated from the plastid to the cytosol [44]. Xanthoxin biosynthesis is a rate-limiting step in ABA biosynthesis, hence NCEDs are major players in the regulation of ABA levels with specific developmental roles. *NCED3* is strongly upregulated by drought stress [45], and in concert with *NCED5* contributes to ABA-mediated drought stress responses [46]. From xanthoxin, bioactive ABA is synthesized in two steps. Firstly, ABA2 converts xanthoxin to abscisic aldehyde [47,48]. Secondly, Arabidopsis aldehyde oxidase 3 (*AAO3*) finally produces ABA from its aldehyde [49,50] in cooperation with a molybdenum cofactor encoded by the Arabidopsis *ABA3* gene [51,52].

Mutants of *aba1* and *aba2* are late-flowering under normal watering conditions [39,43]. Florigen transcript levels are also reduced in these ABA deficient mutants, which is associated with impaired DE compared to the wild type. Notably, the flowering time defect of ABA deficient mutants is restricted to long-day conditions, implying an interaction between ABA production and the photoperiodic response. *ABA2* expression occurs in the phloem companion cells, suggesting that these cells are a major source of ABA production [53]. Phloem-derived ABA may be translocated to other cell types including the shoot via specialized transporters or through the phloematic stream. Thus, while still unknown, the levels and distribution of ABA in the shoot might also affect florigen signaling beyond its site of production.

Insights into the ABA-Flowering Crosstalk from the Analysis of ABA Signaling Mutants

The core ABA signaling cascade is composed of four main proteins and has been excellently reviewed elsewhere [54]. Briefly, ABA is bound by a family of soluble receptors known as PYR/PYL/RCARs [55,56]. Upon binding to ABA, PYR/PYL/RCARs interact with protein phosphatases (PP2C). This interaction inhibits the phosphatase activity of PP2Cs, allowing their substrate, protein kinases of the SNF1-RELATED PROTEIN KINASE 2 (SnRK2) group, to be phosphorylated. Active SnRK2s can, in turn, phosphorylate downstream components including bZIPs encoding ABA-responsive TFs/ABA-responsive element binding factors (ABFs/AREB) [57–59]. Phosphorylated ABFs enact the transcription of ABA/stress-response genes by direct binding on ABA-responsive elements (ABRE) on their promoter sequence [60,61]. In absence of ABA, PP2Cs bind SnRK2s, and keep them in a dephosphorylated, inactive form.

Several ABA signaling genes show expression in the vasculature [43,62]. Other than sharing similar spatial regulation with *FT*, ABA signaling mutants also display flowering defects that are consistent with a role in *FT* activation. Dominant alleles of the *PP2C ABI1* (*abi1-1*) encode proteins that are unable to dissociate from the SnRK2s even in the presence of ABA and thus impair ABA signaling [55,56]. Mutant *abi1-1* plants also fail to activate DE, which is associated with reduced levels of *FT/TSF* transcripts [43]. ABA exogenous applications activate flowering through the ABFs bZIPs which are classified in the same group A of FD-like bZIPs [63]. Interestingly, ABF3 phosphorylation on a LXRXX(S/T) motif, conserved among all ABFs, creates a 14-3-3 binding site [58]. In FD, disruptions in this C-terminal motif prevents FD function [28], and it has been observed in rice that the correct formation of this 14-3-3 binding site is required for the interaction of FD and FT homologs [23].

The *abf2/3/4* triple mutants show large alterations in the ABA-related transcriptome, including deregulation of *PP2C* genes, hinting to the possibility of a transcriptional feedback loop [64]. Additionally, triple *abf2/3/4* and quadruple *abf1/2/3/4* mutants display late flowering phenotypes, with reduced expression of *CO* and its transcriptional activator *FLOWERING BHLH 3* (*FBH3*) [57,65]. Notably, *abf3/4* mutants are late flowering under long-day conditions but not in short-day, and are impaired in DE compared with the wild type. The floral integrator *SUPPRESSOR OF OVEREXPRESSION OF CONSTANS 1* (*SOC1*) is a key target of ABF3 and ABF4 in the leaf. In turn, SOC1 indirectly promotes *FT*—but not *TSF*—expression by negatively regulating a set of *FT* repressors including *TEMPRANILLO1* (*TEM1*), *TEM2*, and *TARGET OF EARLY ACTIVATION TAGGED 1* (*TOE1*) encoding APETALA2 (AP2)-class transcriptional regulators. Interestingly, ABFs bind the *SOC1* promoter through the NF-Y complex by forming a direct interaction with NF-YC subunits. In triple *nf-yc3/4/9* mutants, the DE response is reduced, and *SOC1* transcription is unresponsive to ABA [63]. Beyond their role in the positive regulation of *FT* [17,66], NF-Y TFs are known to be important regulators of ABA-driven transcriptional responses [66]. The wide combinatory range offered by dimerization and trimerization of different NF-Y subunits, each one of these bearing unique functional domains [67], as well as their interaction with other ABA-regulated TFs like the ABFs [63], affects the specificity for DNA targets [68] and provides yet another layer of regulation in the crosstalk between ABA and flowering.

In apparent contrast, it has been reported that water deficit conditions can also repress flowering [39]. This response is observed under short-day conditions when the photoperiodic pathway is inactive. It is hypothesized that the drought-dependent repression of flowering occurs at the shoot meristem, acting independently or downstream of the florigen system. *FLC* plays a major contribution in this process, as mutants of *FLC* do not display delayed flowering in response to water deficit under short-day conditions. Another floral repressor, *SHORT VEGETATIVE PHASE* (*SVP*), a MADS-box type transcriptional regulator structurally related to *FLC* plays a central role in delaying flowering in response to water deficit under short-day conditions. Because FLC and SVP proteins physically interact, it is possible that these similar phenotypes reflect their mode of interaction and targets regulation. *SVP* transcript levels are upregulated in response to water deficit conditions, but not ABA applications [69]. On the other hand, *FLC* transcript levels increase in response to both water deficit and ABA applications [39,70]. Consistent with this ABA-*FLC* regulation, ABA-hypersensitive mutants (derived from loss-of-function alleles of multiple *PP2Cs*) display increased accumulation of *FLC* and are late-flowering compared to the wild type under short-day conditions. *abi1-1* mutants (which are ABA-insensitive) are early-flowering under short days and display reduced levels of *FLC* [43]. The TFs *ABA INSENSITIVE 4* (*ABI4*) encoding an AP2-class protein and the bZIP *ABA INSENSITIVE 5* (*ABI5*) were found to independently target *FLC* to promote its transcriptional activation in response to ABA [70,71].

An important question concerns how ABA levels or signaling may affect the GA-DELLA cascade. Recent data point to a general role for SVP in the control of ABA accumulation in leaves via negative regulation of ABA catabolism pathway genes *CYP707A1*, *CYP707A3*, and *AtBG1*. *svp* mutants display lower cellular ABA contents compared to the wild type and reduced drought stress tolerance [69]. There is also a known contribution of *SVP* at the shoot apex in the control of gibberellic acid biosynthesis, which plays a major role in the activation of flowering under noninductive conditions. SVP acts as a strong repressor of *GIBBERELLIN 20 OXIDASE 2*, encoding an enzyme required for GA biosynthesis. FT triggers the transcriptional repression of *SVP* at the shoot apical meristem (SAM), thus promoting GA accumulation [72]. Thus, variations in *SVP* levels caused by FT or water deficit can affect the ABA–GA balance globally or locally (i.e., in the shoot) to regulate flowering. ABI4 is another node of regulation of the ABA-GA homeostasis by activating the ABA biosynthetic gene *NCED6* and the GA catabolic gene *GIBBERELLIN 2 OXIDASE 7* [73]. ABA and GAs have opposite effects on ABI4 protein accumulation, positive and negative, respectively, indicating that water deficit conditions can alter the ABA–GA balance through modulation of ABI4 cellular abundance. Other than hormone production, the ABA–GA cross talk might occur at the signaling level. Recent studies in tomato indicate that DELLA acts in guard cells to promote stomatal closure, but this effect is ABA-dependent. Moreover, while DELLA in guard cells does not affect ABA levels, it increases guard cell ABA responsiveness [74]. While it is unknown whether this model can apply to Arabidopsis, it points to alternative modes of ABA–GA cross regulations possibly occurring at different tissue scales.

3. ABA Signaling Integration through GIGANTEA

GIGANTEA (*GI*) was identified as a key flowering gene, required for photoperiod perception and clock function. *GI* is also emerging as the key driver of DE, independent of its known role in the photoperiodic cascade. Given the multiple regulatory mechanisms coordinated by GI in the flowering regulatory process, it would be relevant to understand which step(s) could be sensitive to ABA levels. Here, we shall focus on the emerging role of GI in mediating hormonal signals (emphasizing the link to ABA signaling) and refer the reader to recent reviews detailing the mechanism of GI in photoperiodic and clock regulation [75].

Genetic evidence indicates that GI function is sensitive to ABA signaling status [43]. Impairing ABA signaling (as in *abi1-1*—mutants) causes marked reductions in *FT* and *TSF* accumulation even in a genetic background where *GI* is expressed constitutively via the *35S* promoter. Interestingly, *CO* levels are only moderately reduced in *35S::GI abi1-1* plants compared to *35S::GI*. This result supports a model where some aspects of GI protein function important for *FT*—but not *CO*—transcriptional regulation are sensitive to ABA signaling.

The idea that GI relay ABA signals onto *FT* with minor contributions from CO also derives from the study of *cdf1/2/3/5/gi* quintuple mutants characterized by high levels of *CO* transcript and an early flowering phenotype. These mutants failed to upregulate *FT* under water deficit conditions (as compared to *cdf1/2/3/5* quadruple mutants) and to activate DE, supporting the pivotal role of florigens in this process [43]. The interpretation of these results is that GI protein function is required to confer ABA-dependent responsiveness at the *FT* promoter.

The precise mode of GI-dependent florigen regulation promoted by ABA is still unclear. Drought or ABA alone cannot activate *FT* expression in *co* mutants and indicates that an interplay between GI and CO is ultimately necessary for *FT* activation and DE to occur. Interestingly, the florigen *TSF* can be transcriptionally activated in *co* mutants under water deficit conditions in a GI-dependent manner, indicating that in some cases, the interplay between GI and ABA is sufficient in promoting florigen expression [43]. This observation echoes the results of misexpression studies showing that GI can directly activate *FT* in the vasculature and partially rescue the late-flowering phenotype of *co* mutants [76]. GI is enriched at the *FT* promoter region, at positions usually occupied by strong *FT* repressors including SVP and the aforementioned TEM1/2, which are negatively regulated via the *ABF/SOC1* axis. The CDFs are also repressors of *FT*, and GI is required to relieve their repression at the promoter of *FT* [77] (Figure 1). Because GI does not present an obvious DNA binding domain, it may be recruited at the *FT* promoter through independent protein–protein interaction events to facilitate chromatin accessibility of positive regulators.

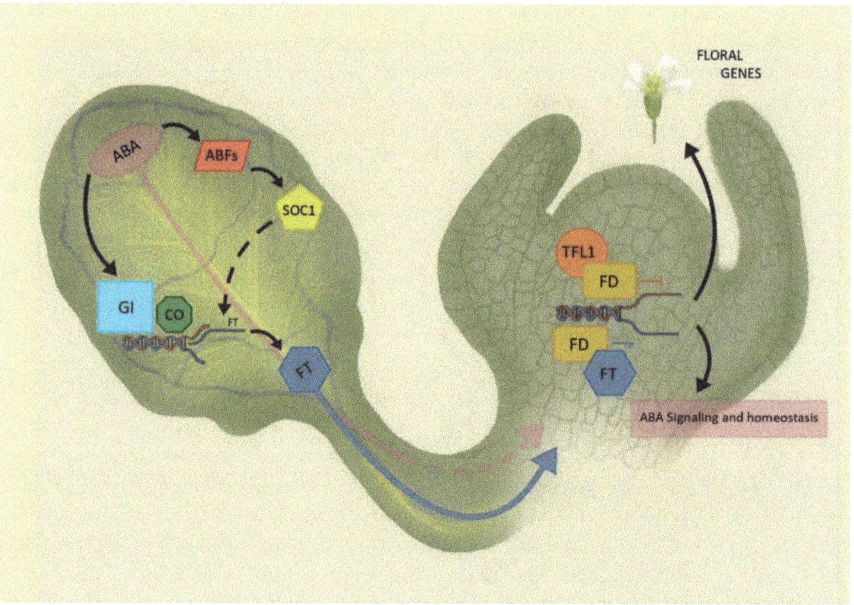

Figure 1. Abscisic acid (ABA) signaling and flowering regulation. In the leaves (left), ABA controls FLOWERING LOCUS T (FT) transcription acting on GIGANTEA (GI) and CONSTANS (CO); ABA-responsive transcription factors (ABFs) can modulate SUPPRESSOR OF OVEREXPRESSION OF CONSTANS 1 (SOC1) expression, in turn affecting FT transcription through an indirect mechanism. FT moves to the SAM where it interacts with FD and FD-like basic leucine zippers (bZIPs) to activate floral genes and ABA signaling transcriptome. ABA is transported in the phloem, but its roles at the SAM are not yet known. TERMINAL FLOWER 1 (TFL1) antagonizes FT, repressing transcription. Dashed lines represent indirect or not yet confirmed pathways, while full lines represent known ones.

Chromatin immunoprecipitation (ChIP), followed by sequencing experiments, revealed that GI is recruited to the chromatin of thousands of loci to regulate gene expression. Interestingly, GI ChIP-seq peaks occur on regulatory regions of ABA/water deficit-related genes which are also differentially expressed genes in *gi* mutants [78]. GI was shown to interact with the ABA-related bZIP DC3 PROMOTER-BINDING FACTOR 4/ENHANCED EM LEVEL (DPBF4/EEL) to activate drought responses through the activation of the ABA biosynthetic gene *NCED3*. *eel*, *gi-1* and the corresponding double mutant have significantly lower expression of *NCED3* and present impaired stomatal closure in response to dehydration, thereby displaying a low survival rate in water deficit conditions [79]. These results could lead to a model where ABA-activated TFs recruit GI at different genomic positions to regulate gene networks related to drought stress. GI is found in complex with various enzymatic functions including kinases [80], O-fucosyltransferases [81], HEAT SHOCK PROTEIN 70/90 co-chaperones that promote the maturation of client protein interactors [82]. The most recent updates about potential and confirmed GI interactions can be found in [83]. The ever-growing list of interactors may suggest that GI acts as a scaffold protein, providing different enzymatic activities, possibly in conjunction with its recruitment at different genomic positions. Because GI protein levels oscillate during the day in a circadian manner, its recruitment to chromatin may gate DNA accessibility to TFs, and thus coordinate plant sensitivity to external signals in a diel manner [78]. Knowledge of these transcriptional mechanisms may help understand how GI can influence such a vast array of cellular responses.

4. Role of FD and FD-Like bZIPs Protein Complexes in Modulating ABA Signaling

An important theme arising from recent studies concerns the putative role of FD and FD-like proteins in the modulation of ABA signaling in complex with FT-like proteins. The Arabidopsis genome encodes 78 bZIP TFs, classified into 13 groups. These bZIPs have a basic domain required for the DNA binding activity and a characteristic leucine zipper domain that allows for homo- and heterodimerization [84]. The key floral genes *FD* and *FD PARALOGUE* (*FDP*) belong to group A of Arabidopsis bZIPs, totaling 13 members. Consistent with its established role in flowering, *FD* controls the expression of floral regulators like *SOC1*, *LEAFY*, *FRUITFULL*, and *APETALA1* [24,85–87] by direct binding at their respective promoters [87]. FD and FDP also share several downstream targets that are ABA- and water stress-related, including other members of group A bZIPs such as *ABF3* and *DPBF1/ABI5*, the *PP2Cs ABI1* and *HAB1*, the ABA catabolic gene *CYP707A2*, and proteostasis-related genes *ABI FIVE BINDING PROTEIN 2* (*AFP2*) and *AFP4* (Figure 1). *fd* and *fdp* mutant seedlings display reduced ABA sensitivity in germination assays [87]. Hence, one emerging aspect related to FD and FDP function is their role in the control of ABA response and metabolism. While the organization of these regulatory networks at the shoot apex is currently unclear, further confirmation for this interplay between FD and ABA-related genes derives from ChIP and expression analyses of the shoot-specific *TFL1* gene. The TFL1-FD complex directly represses, among others, the ABA biosynthetic gene *ABA1*, the bZIPs *ABF4/AREB2* and *ABI5* together with the *ABI5* regulator *AFP2* [27]. Thus, similar to floral targets, a competition between FT and TFL1 might modulate ABA levels or sensitivity in the shoot meristem cells (Figure 1).

During seed development, TFL1 was shown to stabilize ABI5 protein in the developing endosperm, possibly in response to ABA [88]. ABA involvement in seed development is known [89]. However, the newly discovered TFL1–ABI5 interaction further indicates multiple regulatory FD-like bZIP complexes that might have different roles and functions according to the tissue and developmental specific context in which these complexes form. ABI5 was described as a floral repressor, with transgenic plants overexpressing ABI5 showing delayed flowering under long-day conditions, owing to increased levels of *FLC* [70]. With the notable exclusion of FD and FDP, most of the molecular events involving the Arabidopsis group A bZIPs were studied in seedlings, and little is known about their targets at the SAM during floral transition. However, many of the characteristics of these ABA related bZIPs—notably, their ability to heterodimerize, to bind to 14-3-3 proteins, and their structural similarity with FD and FDP—could hint to a more prominent role for this protein family at the apex. It is possible to hypothesize a highly fluid and dynamic model acting at the SAM. Different bZIPs can be activated by the relocation of FT and TSF to the SAM, and act in cooperation or antagonistically with FD and FDP, to integrate photoperiodic and hormonal signals (Figure 1). Different combinations of group A bZIP heterodimers could further allow or deny protein–protein interactions [90], and ultimately target different subsets of downstream components of flowering or ABA-related signal cascades (Figure 1).

5. Molecular Insights into the ABA–Flowering Relationship in Crops

While crop production is facing the threat of climate change, with extreme meteorological drought predicted to be more common [91], DE response could either be either beneficial or maladaptive, depending on the drought scenario [92]. Maintaining a prolonged vegetative growth could give a competitive advantage in terms of seed number, while DE and a short life cycle can increase reproductive success in drought-prone environments at the expense of productivity and yield under sufficient watering conditions [93,94]. Taking into account our continuous progresses in deciphering the ABA–flowering molecular interactions in Arabidopsis, a key question is how this knowledge can be translated to other species, including crops. Valuable information may derive from the study of DE traits in natural populations as shifts in flowering time phenology is a major trait enabling their survival [95,96]. This wealth of knowledge will lead into molecularly exposed allelic variations that confer plasticity in DE and can help uncoupling flowering responses from generic drought and abiotic stress responses, providing novel breeding and biotechnological targets for crop improvement.

In the model monocot crop rice (*Oryza sativa* L.), domesticated at tropical latitudes, flowering is induced by the transition from long to short day conditions [97], while in temperate rice varieties, flowering is photoperiod-insensitive [98]. Interestingly, the DE pathway is conserved in its essential components in rice. ABA is required to activate the DE pathway in rice, with ABA-deficient rice mutants being impaired in the DE response. *OsGIGANTEA* (*OsGI*) is involved in the DE response, and *OsGI-RNAi* lines present reduced DE response, albeit its role appears to be ABA-independent [99]. In apparent contrast, severe drought stress delays flowering in rice, with the repression of the rice florigens *HEADING DATE 3A* (*Hd3a*) and *RICE FLOWERING LOCUS T 1* (*RFT1*) [100]. Delaying flowering under drought stress could also be mediated by the rice floral repressor *RICE CENTRORADIALIS 1*, a TFL-like gene induced by ABA under severe stress conditions [101]. Following a biotechnological approach, Miao et al. [102] used gene editing techniques to obtain rice mutants that are less sensitive to ABA by mutating multiple genes of the ABA receptor *PYR/PYL/RCAR* family. These lines had higher growth rates and improved yields compared to those of the wild type (~+25%), but also displayed increased sensitivity to water stress conditions. Higher productivity was in part determined by an extended duration of the vegetative phase. Unlike *OsPYL* loss-of-function mutants, overexpression of *OsPYL/RCAR5* improved salt and drought tolerance in rice during the vegetative growth stage. In contrast, in normal watering conditions, seed yield was greatly reduced (~−75%) [103], pointing to a tradeoff between growth duration and drought tolerance traits. Studies with introgression lines led to the discovery that most drought tolerance-associated quantitative trait loci (QTLs) are independent of DE-QTLs, pointing at the evolution of at least two distinct adaptive strategies in rice under drought stress [104]. DE seems an effective strategy that might contribute to improving yield under stress. However, this contribution varies depending on specific drought scenarios, and on the developmental stage in which drought stress is imposed [105]. Collectively, these results imply a significant and yet uncharacterized contribution of ABA in the control of the floral transition of rice, whereby DE activation depends on the genotype and the intensity and timing of the drought stress.

Little information is available about ABA molecular control of flowering in other monocots, however, in a pioneeristic attempt to produce drought-resistant crops, transgenic maize plants constitutively expressing *ZmNF-YB2* showed a range of ABA-related developmental responses, including higher stomatal conductance and chlorophyll content and delayed onset of senescence, leading to improved yield under severe drought conditions. These advantages under drought conditions were highly situational [106]. Fast cycling, early flowering crops could avoid terminal drought and reduce the length of the crop season, but often this strategy pays a cost in terms of reduced yield [107,108]. It is still unclear what genetic adjustment may be needed to manipulate ABA sensitivity and flowering time. Resolving the flowering-specific effects of ABA from its general role in drought stress response may lead to improvements to crop yield whilst maintaining stress responsiveness in specific environments.

Author Contributions: D.M. and L.C. outlined and wrote the manuscript. B.S. wrote Section 3 together with L.C. A.L. wrote Section 5 together with L.C.; L.C. acquired funding. All authors reviewed and edited the manuscript. All authors have read and agreed to the published version of the manuscript.

Funding: This work was mainly supported by a Research Grant from HFSP (Ref.-No: RGP0011/2019) and in part by the Linea 3—SEED 2019—DISENGAGE funded by the University of Milan.

Acknowledgments: A special thanks to Erika Pagliari (erikapagliari.ekp@gmail.com) for the artwork and graphics of Figure 1.

Conflicts of Interest: The authors declare no conflict of interests.

Abbreviations

ABA	Abscisic acid
ABFs	ABA-responsive TFs/ABA-responsive element binding factors
bZIP	basic leucine zipper
CDFs	CYCLING DOF FACTORs
CO	CONSTANS
CORE	CONSTANS RESPONSIVE elements
DE	Drought Escape
DPBFs	Dc3 promoter-binding factors
FKF1	FLAVIN-BINDING, KELCH REPEAT, F-BOX1
FLC	FLOWERING LOCUS C
FRI	FRIGIDA
FT	FLOWERING LOCUS T
GA	Gibberellic acid, gibberellins
GI	GIGANTEA
NF-Y	Nuclear Factor Y
QTLs	Quantitative Trait Loci
SOC1	SUPPRESSOR OF OVEREXPRESSION OF CONSTANS 1
TFL1	TERMINAL FLOWER 1
TFs	Transcription factors
TSF	TWIN SISTER OF FT

References

1. Wolters, H.; Jürgens, G. Survival of the flexible: Hormonal growth control and adaptation in plant development. *Nat. Rev. Genet.* **2009**, *10*, 305–317. [CrossRef] [PubMed]
2. Zhang, D.P. *Abscisic Acid: Metabolism, Transport and Signaling*, 1st ed.; Springer: Heidelberg, Germany, 2014.
3. Li, S.; Li, X.; Wei, Z.; Liu, F. ABA-mediated modulation of elevated CO_2 on stomatal response to drought. *Curr. Opin. Plant Biol.* **2020**, *56*, 174–180. [CrossRef] [PubMed]
4. Yoshida, T.; Christmann, A.; Yamaguchi-Shinozaki, K.; Grill, E.; Fernie, A.R. Revisiting the Basal Role of ABA—Roles Outside of Stress. *Trends Plant Sci.* **2019**, *24*, 625–635. [CrossRef] [PubMed]
5. Martínez-Zapater, J.M.; Coupland, G.; Dean, C.; Koornneef, M. The Transition to Flowering in Arabidopsis. In *Cold Spring Harbor Monograph Archive*; Cold Spring Harbor Laboratory Press: New York, NY, USA, 1994; Volume 27, pp. 403–433.
6. Kinoshita, A.; Richter, R. Genetic and molecular basis of floral induction in *Arabidopsis thaliana*. *J. Exp. Bot.* **2020**, *71*, 2490–2504. [CrossRef] [PubMed]
7. Song, Y.H.; Shim, J.S.; Kinmonth-Schultz, H.A.; Imaizumi, T. Photoperiodic Flowering: Time Measurement Mechanisms in Leaves. *Annu. Rev. Plant Biol.* **2015**, *66*, 441–464. [CrossRef] [PubMed]
8. An, H.; Roussot, C.; Suárez-López, P.; Corbesier, L.; Vincent, C.; Piñeiro, M.; Hepworth, S.; Mouradov, A.; Justin, S.; Turnbull, C.; et al. CONSTANS acts in the phloem to regulate a systemic signal that induces photoperiodic flowering of *Arabidopsis*. *Development* **2004**, *131*, 3615–3626. [CrossRef] [PubMed]

9.   Liu, L.; Liu, C.; Hou, X.; Xi, W.; Shen, L.; Tao, Z.; Wang, Y.; Yu, H. FTIP1 Is an Essential Regulator Required for Florigen Transport. *PLoS Biol.* **2012**, *10*, e1001313. [CrossRef]

10. Jaeger, K.E.; Wigge, P.A. FT Protein Acts as a Long-Range Signal in *Arabidopsis*. *Curr. Biol.* **2007**, *17*, 1050–1054. [CrossRef]

11. Corbesier, L.; Vincent, C.; Jang, S.; Fornara, F.; Fan, Q.; Searle, I.; Giakountis, A.; Farrona, S.; Gissot, L.; Turnbull, C.; et al. FT protein movement contributes to long-distance signaling in floral induction of *Arabidopsis*. *Science* **2007**, *316*, 1030–1033. [CrossRef]

12. Mathieu, J.; Warthmann, N.; Küttner, F.; Schmid, M. Export of FT Protein from Phloem Companion Cells Is Sufficient for Floral Induction in Arabidopsis. *Curr. Biol.* **2007**, *17*, 1055–1060. [CrossRef]

13. Shim, J.S.; Kubota, A.; Imaizumi, T. Circadian clock and photoperiodic flowering in *Arabidopsis*: CONSTANS is a Hub for Signal integration. *Plant Physiol.* **2017**, *173*, 5–15. [CrossRef] [PubMed]

14. Imaizumi, T.; Tran, H.G.; Swartz, T.E.; Briggs, W.R.; Kay, S.A. FKF1 is essential for photoperiodic-specific light signalling in *Arabidopsis*. *Nature* **2003**, *426*, 302–306. [CrossRef] [PubMed]

15. Fornara, F.; Panigrahi, K.C.S.; Gissot, L.; Sauerbrunn, N.; Rühl, M.; Jarillo, J.A.; Coupland, G. *Arabidopsis* DOF Transcription Factors Act Redundantly to Reduce CONSTANS Expression and Are Essential for a Photoperiodic Flowering Response. *Dev. Cell* **2009**, *17*, 75–86. [CrossRef] [PubMed]

16. Suárez-López, P.; Wheatley, K.; Robson, F.; Onouchi, H.; Valverde, F.; Coupland, G. CONSTANS mediates between the circadian clock and the control of flowering in *Arabidopsis*. *Nature* **2001**, *410*, 1116–1120. [CrossRef]

17. Gnesutta, N.; Mantovani, R.; Fornara, F. Plant Flowering: Imposing DNA Specificity on Histone-Fold Subunits. *Trends Plant Sci.* **2018**, *23*, 293–301. [CrossRef]

18. Chaves-Sanjuan, A.; Gnesutta, N.; Gobbini, A.; Martignago, D.; Bernardini, A.; Fornara, F.; Mantovani, R.; Nardini, M. Structural determinants for NF-Y subunit organization and NF-Y/DNA association in plants. *Plant J.* **2020**. [CrossRef]

19. Brambilla, V.; Fornara, F. Y flowering? Regulation and activity of CONSTANS and CCT-domain proteins in *Arabidopsis* and crop species. *Biochim. Biophys. Acta Gene Regul. Mech.* **2017**, *1860*, 655–660. [CrossRef]

20. Golembeski, G.S.; Imaizumi, T. Photoperiodic Regulation of Florigen Function in *Arabidopsis thaliana*. *Arab. Book* **2015**, *13*, e0178. [CrossRef]

21. Wang, Y.; Gu, X.; Yuan, W.; Schmitz, R.J.; He, Y. Photoperiodic control of the floral transition through a distinct polycomb repressive complex. *Dev. Cell* **2014**. [CrossRef]

22. Pin, P.A.; Nilsson, O. The multifaceted roles of FLOWERING LOCUS T in plant development. *Plant Cell Environ.* **2012**, *35*, 1742–1755. [CrossRef]

23. Taoka, K.; Ohki, I.; Tsuji, H.; Furuita, K.; Hayashi, K.; Yanase, T.; Yamaguchi, M.; Nakashima, C.; Purwestri, Y.A.; Tamaki, S.; et al. 14-3-3 proteins act as intracellular receptors for rice Hd3a florigen. *Nature* **2011**, *476*, 332–335. [CrossRef] [PubMed]

24. Collani, S.; Neumann, M.; Yant, L.; Schmid, M. FT modulates genome-wide DNA-binding of the bZIP transcription factor FD. *Plant Physiol.* **2019**, *180*, 367–380. [CrossRef] [PubMed]

25. Conti, L.; Bradley, D. TERMINAL FLOWER1 is a mobile signal controlling *Arabidopsis* architecture. *Plant Cell* **2007**, *19*, 767–778. [CrossRef] [PubMed]

26. Wickland, D.P.; Hanzawa, Y.; Abe, M.; Kobayashi, Y.; Yamamoto, S.; Daimon, Y.; Yamaguchi, A.; Ikeda, Y.; Ichinoki, H.; Notaguchi, M.; et al. The FLOWERING LOCUS T/TERMINAL FLOWER 1 Gene Family: Functional Evolution and Molecular Mechanisms. *Mol. Plant* **2015**. [CrossRef]

27. Zhu, Y.; Klasfeld, S.; Jeong, C.W.; Jin, R.; Goto, K.; Yamaguchi, N.; Wagner, D. TERMINAL FLOWER 1-FD complex target genes and competition with FLOWERING LOCUS T. *Nat. Commun.* **2020**, *11*, 5118. [CrossRef]

28. Abe, M.; Kobayashi, Y.; Yamamoto, S.; Daimon, Y.; Yamaguchi, A.; Ikeda, Y.; Ichinoki, H.; Notaguchi, M.; Goto, K.; Araki, T. FD, a bZIP protein mediating signals from the floral pathway integrator FT at the shoot apex. *Science* **2005**, *309*, 1052–1056. [CrossRef]

29. Goretti, D.; Silvestre, M.; Collani, S.; Langenecker, T.; Méndez, C.; Madueño, F.; Schmid, M. TERMINAL FLOWER1 functions as a mobile transcriptional cofactor in the shoot apical meristem. *Plant Physiol.* **2020**, *182*, 2081–2095. [CrossRef]

30. Zhu, D.; Rosa, S.; Dean, C. Nuclear organization changes and the epigenetic silencing of FLC during vernalization. *J. Mol. Biol.* **2015**, *427*, 659–669. [CrossRef]

31. Hepworth, J.; Dean, C. Flowering locus C's lessons: Conserved chromatin switches underpinning developmental timing and adaptation. *Plant Physiol.* **2015**, *168*, 1237–1245. [CrossRef]

32. Searle, I.; He, Y.; Turck, F.; Vincent, C.; Fornara, F.; Kröber, S.; Amasino, R.A.; Coupland, G. The transcription factor FLC confers a flowering response to vernalization by repressing meristem competence and systemic signaling in *Arabidopsis*. *Genes Dev.* **2006**, *20*, 898–912. [CrossRef]

33. Bao, S.; Hua, C.; Shen, L.; Yu, H. New insights into gibberellin signaling in regulating flowering in *Arabidopsis*. *J. Integr. Plant Biol.* **2020**, *62*, 118–131. [CrossRef] [PubMed]

34. Mutasa-Göttgens, E.; Hedden, P. Gibberellin as a factor in floral regulatory networks. *J. Exp. Bot.* **2009**, *60*, 1979–1989. [CrossRef] [PubMed]

35. Wilson, R.N.; Heckman, J.W.; Somerville, C.R. Gibberellin is required for flowering in *Arabidopsis thaliana* under short days. *Plant Physiol.* **1992**, *100*, 403–408. [CrossRef] [PubMed]

36. Davière, J.M.; Achard, P. A Pivotal Role of DELLAs in Regulating Multiple Hormone Signals. *Mol. Plant* **2016**, *9*, 10–20. [CrossRef]

37. Van De Velde, K.; Ruelens, P.; Geuten, K.; Rohde, A.; Van Der Straeten, D. Exploiting DELLA Signaling in Cereals. *Trends Plant Sci.* **2017**, *22*, 880–893. [CrossRef]

38. Izawa, T. What is going on with the hormonal control of flowering in plants? *Plant J.* **2020**. [CrossRef]

39. Riboni, M.; Galbiati, M.; Tonelli, C.; Conti, L. GIGANTEA enables drought escape response via abscisic acid-dependent activation of the florigens and SUPPRESSOR OF OVEREXPRESSION OF CONSTANS. *Plant Physiol.* **2013**, *162*, 1706–1719. [CrossRef]

40. Kooyers, N.J. The evolution of drought escape and avoidance in natural herbaceous populations. *Plant Sci.* **2015**, *234*, 155–162. [CrossRef]

41. Juenger, T.E. Natural variation and genetic constraints on drought tolerance. *Curr. Opin. Plant Biol.* **2013**, *16*, 274–281. [CrossRef]

42. Han, Y.; Zhang, X.; Wang, Y.; Ming, F. The suppression of WRKY44 by GIGANTEA-miR172 pathway is involved in drought response of *Arabidopsis thaliana*. *PLoS ONE* **2013**, *8*, e73541. [CrossRef]

43. Riboni, M.; Robustelli Test, A.; Galbiati, M.; Tonelli, C.; Conti, L. ABA-dependent control of GIGANTEA signalling enables drought escape via up-regulation of FLOWERING LOCUS T in *Arabidopsis thaliana*. *J. Exp. Bot.* **2016**, *67*, 6309–6322. [CrossRef] [PubMed]

44. Schwartz, S.H.; Tan, B.C.; McCarty, D.R.; Welch, W.; Zeevaart, J.A.D. Substrate specificity and kinetics for VP14, a carotenoid cleavage dioxygenase in the ABA biosynthetic pathway. *Biochim. Biophys. Acta Gen. Subj.* **2003**, *1619*, 9–14. [CrossRef]

45. Iuchi, S.; Kobayashi, M.; Taji, T.; Naramoto, M.; Seki, M.; Kato, T.; Tabata, S.; Kakubari, Y.; Yamaguchi-Shinozaki, K.; Shinozaki, K. Regulation of drought tolerance by gene manipulation of 9-cis-epoxycarotenoid dioxygenase, a key enzyme in abscisic acid biosynthesis in *Arabidopsis*. *Plant J.* **2001**, *27*, 325–333. [CrossRef] [PubMed]

46. Frey, A.; Effroy, D.; Lefebvre, V.; Seo, M.; Perreau, F.; Berger, A.; Sechet, J.; To, A.; North, H.M.; Marion-Poll, A. Epoxycarotenoid cleavage by NCED5 fine-tunes ABA accumulation and affects seed dormancy and drought tolerance with other NCED family members. *Plant J.* **2012**, *70*, 501–512. [CrossRef] [PubMed]

47. Cheng, W.H.; Endo, A.; Zhou, L.; Penney, J.; Chen, H.C.; Arroyo, A.; Leon, P.; Nambara, E.; Asami, T.; Seo, M.; et al. A unique short-chain dehydrogenase/reductase in arabidopsis glucose signaling and abscisic acid biosynthesis and functions. *Plant Cell* **2002**, *14*, 2723–2743. [CrossRef]

48. González-Guzmán, M.; Apostolova, N.; Bellés, J.M.; Barrero, J.M.; Piqueras, P.; Ponce, M.R.; Micol, J.L.; Serrano, R.; Rodríguez, P.L. The short-chain alcohol dehydrogenase ABA2 catalyzes the conversion of xanthoxin to abscisic aldehyde. *Plant Cell* **2002**, *14*, 1833–1846. [CrossRef]

49. Seo, M.; Peeters, A.J.M.; Koiwai, H.; Oritani, T.; Marion-Poll, A.; Zeevaart, J.A.D.; Koornneef, M.; Kamiya, Y.; Koshiba, T. The *Arabidopsis* aldehyde oxidase 3 (AAO3) gene product catalyzes the final step in abscisic acid biosynthesis in leaves. *Proc. Natl. Acad. Sci. USA* **2000**, *97*, 12908–12913. [CrossRef]

50. Seo, M.; Aoki, H.; Koiwai, H.; Kamiya, Y.; Nambara, E.; Koshiba, T. Comparative studies on the *Arabidopsis* aldehyde oxidase (AAO) gene family revealed a major role of AAO3 in ABA biosynthesis in seeds. *Plant Cell Physiol.* **2004**, *45*, 1694–1703. [CrossRef]

51. Bittner, F.; Oreb, M.; Mendel, R.R. ABA3 Is a Molybdenum Cofactor Sulfurase Required for Activation of Aldehyde Oxidase and Xanthine Dehydrogenase in Arabidopsis thaliana. *J. Biol. Chem.* **2001**, *276*, 40381–40384. [CrossRef]

52. Xiong, L.; Ishitani, M.; Lee, H.; Zhu, J.K. The Arabidopsis LOS5/ABA3 locus encodes a molybdenum cofactor sulfurase and modulates cold stress- and osmotic stress-responsive gene expression. *Plant Cell* **2001**, *13*, 2063–2083. [CrossRef]

53. Kuromori, T.; Sugimoto, E.; Shinozaki, K. Intertissue signal transfer of abscisic acid from vascular cells to guard cells. *Plant Physiol.* **2014**, *164*, 1587–1592. [CrossRef] [PubMed]

54. Umezawa, T.; Nakashima, K.; Miyakawa, T.; Kuromori, T.; Tanokura, M.; Shinozaki, K.; Yamaguchi-Shinozaki, K. Molecular basis of the core regulatory network in ABA responses: Sensing, signaling and transport. *Plant Cell Physiol.* **2010**, *51*, 1821–1839. [CrossRef] [PubMed]

55. Ma, Y.; Szostkiewicz, I.; Korte, A.; Moes, D.; Yang, Y.; Christmann, A.; Grill, E. Regulators of PP2C phosphatase activity function as abscisic acid sensors. *Science* **2009**, *324*, 1064–1068. [CrossRef] [PubMed]

56. Park, S.Y.; Fung, P.; Nishimura, N.; Jensen, D.R.; Fujii, H.; Zhao, Y.; Lumba, S.; Santiago, J.; Rodrigues, A.; Chow, T.F.F.; et al. Abscisic acid inhibits type 2C protein phosphatases via the PYR/PYL family of START proteins. *Science* **2009**, *324*, 1068–1071. [CrossRef] [PubMed]

57. Yoshida, T.; Fujita, Y.; Maruyama, K.; Mogami, J.; Todaka, D.; Shinozaki, K.; Yamaguchi-Shinozaki, K. Four *Arabidopsis* AREB/ABF transcription factors function predominantly in gene expression downstream of SnRK2 kinases in abscisic acid signalling in response to osmotic stress. *Plant Cell Environ.* **2015**, *38*, 35–49. [CrossRef] [PubMed]

58. Sirichandra, C.; Davanture, M.; Turk, B.E.; Zivy, M.; Valot, B.; Leung, J.; Merlot, S. The arabidopsis ABA-activated kinase OST1 phosphorylates the bZIP transcription factor ABF3 and creates a 14-3-3 binding site involved in its turnover. *PLoS ONE* **2010**, *5*, e13935. [CrossRef]

59. Furihata, T.; Maruyama, K.; Fujita, Y.; Umezawa, T.; Yoshida, R.; Shinozaki, K.; Yamaguchi-Shinozaki, K. Abscisic acid-dependent multisite phosphorylation regulates the activity of a transcription activator AREB1. *Proc. Natl. Acad. Sci. USA* **2006**, *103*, 1988–1993. [CrossRef]

60. Yoshida, T.; Mogami, J.; Yamaguchi-Shinozaki, K. Omics approaches toward defining the comprehensive abscisic acid signaling network in plants. *Plant Cell Physiol.* **2014**, *56*, 1043–1052. [CrossRef]

61. Nakashima, K.; Yamaguchi-Shinozaki, K. ABA signaling in stress-response and seed development. *Plant Cell Rep.* **2013**, *32*, 959–970. [CrossRef]

62. Gonzalez-Guzman, M.; Pizzio, G.A.; Antoni, R.; Vera-Sirera, F.; Merilo, E.; Bassel, G.W.; Fernández, M.A.; Holdsworth, M.J.; Perez-Amador, M.A.; Kollist, H.; et al. Arabidopsis PYR/PYL/RCAR receptors play a major role in quantitative regulation of stomatal aperture and transcriptional response to abscisic acid. *Plant Cell* **2012**, *24*, 2483–2496. [CrossRef]

63. Hwang, K.; Susila, H.; Nasim, Z.; Jung, J.Y.; Ahn, J.H. Arabidopsis ABF3 and ABF4 Transcription Factors Act with the NF-YC Complex to Regulate SOC1 Expression and Mediate Drought-Accelerated Flowering. *Mol. Plant* **2019**, *12*, 489–505. [CrossRef] [PubMed]

64. Yoshida, T.; Fujita, Y.; Sayama, H.; Kidokoro, S.; Maruyama, K.; Mizoi, J.; Shinozaki, K.; Yamaguchi-Shinozaki, K. AREB1, AREB2, and ABF3 are master transcription factors that cooperatively regulate ABRE-dependent ABA signaling involved in drought stress tolerance and require ABA for full activation. *Plant J.* **2010**, *61*, 672–685. [CrossRef] [PubMed]

65. Siriwardana, C.L.; Gnesutta, N.; Kumimoto, R.W.; Jones, D.S.; Myers, Z.A.; Mantovani, R.; Holt, B.F. NUCLEAR FACTOR Y, Subunit A (NF-YA) Proteins Positively Regulate Flowering and Act Through FLOWERING LOCUS T. *PLoS Genet.* **2016**, *12*, e1006496. [CrossRef] [PubMed]

66. Song, L.; Huang, S.S.C.; Wise, A.; Castanoz, R.; Nery, J.R.; Chen, H.; Watanabe, M.; Thomas, J.; Bar-Joseph, Z.; Ecker, J.R. A transcription factor hierarchy defines an environmental stress response network. *Science* **2016**, *354*. [CrossRef]

67. Gnesutta, N.; Saad, D.; Chaves-Sanjuan, A.; Mantovani, R.; Nardini, M. Crystal Structure of the *Arabidopsis thaliana* L1L/NF-YC3 Histone-fold Dimer Reveals Specificities of the LEC1 Family of NF-Y Subunits in Plants. *Mol. Plant* **2017**, *10*, 645–648. [CrossRef]

68. Gnesutta, N.; Chiara, M.; Bernardini, A.; Balestra, M.; Horner, D.S.; Mantovani, R. The plant NF-Y DNA matrix in vitro and in vivo. *Plants* **2019**, *8*, 406. [CrossRef]

69. Wang, Z.; Wang, F.; Hong, Y.; Yao, J.; Ren, Z.; Shi, H.; Zhu, J.K. The Flowering Repressor SVP Confers Drought Resistance in Arabidopsis by Regulating Abscisic Acid Catabolism. *Mol. Plant* **2018**, *11*, 1184–1197. [CrossRef]

70. Wang, Y.; Li, L.; Ye, T.; Lu, Y.; Chen, X.; Wu, Y. The inhibitory effect of ABA on floral transition is mediated by ABI5 in *Arabidopsis*. *J. Exp. Bot.* **2013**, *64*, 675–684. [CrossRef]

71. Shu, K.; Chen, Q.; Wu, Y.; Liu, R.; Zhang, H.; Wang, S.; Tang, S.; Yang, W.; Xie, Q. ABSCISIC ACID-INSENSITIVE 4 negatively regulates flowering through directly promoting Arabidopsis FLOWERING LOCUS C transcription. *J. Exp. Bot.* **2016**, *67*, 195–205. [CrossRef]

72. Porri, A.; Torti, S.; Mateos, J.; Romera-Branchat, M.; García-Martínez, J.L.; Fornara, F.; Gregis, V.; Kater, M.M.; Coupland, G. SHORT VEGETATIVE PHASE reduces gibberellin biosynthesis at the *Arabidopsis* shoot apex to regulate the floral transition Fernando Andrés1. *Proc. Natl. Acad. Sci. USA* **2014**, *111*, E2760–E2769. [CrossRef]

73. Shu, K.; Chen, Q.; Wu, Y.; Liu, R.; Zhang, H.; Wang, P.; Li, Y.; Wang, S.; Tang, S.; Liu, C.; et al. ABI4 mediates antagonistic effects of abscisic acid and gibberellins at transcript and protein levels. *Plant J.* **2016**, *85*, 348–361. [CrossRef] [PubMed]

74. Nir, I.; Shohat, H.; Panizel, I.; Olszewski, N.; Aharoni, A.; Weiss, D. The tomato DELLA protein PROCERA acts in guard cells to promote stomatal closure. *Plant Cell* **2017**, *29*, 3186–3197. [CrossRef] [PubMed]

75. Mishra, P.; Panigrahi, K.C. GIGANTEA—An emerging story. *Front. Plant Sci.* **2015**, *6*. [CrossRef] [PubMed]

76. Sawa, M.; Kay, S.A. GIGANTEA directly activates Flowering Locus T in *Arabidopsis thaliana*. *Proc. Natl. Acad. Sci. USA* **2011**, *108*, 11698–11703. [CrossRef] [PubMed]

77. Song, Y.H.; Smith, R.W.; To, B.J.; Millar, A.J.; Imaizumi, T. FKF1 conveys timing information for CONSTANS stabilization in photoperiodic flowering. *Science* **2012**, *336*, 1045–1049. [CrossRef]

78. Nohales, M.A.; Liu, W.; Duffy, T.; Nozue, K.; Sawa, M.; Pruneda-Paz, J.L.; Maloof, J.N.; Jacobsen, S.E.; Kay, S.A. Multi-level Modulation of Light Signaling by GIGANTEA Regulates Both the Output and Pace of the Circadian Clock. *Dev. Cell* **2019**, *49*, 840–851. [CrossRef]

79. Baek, D.; Kim, W.Y.; Cha, J.Y.; Park, H.J.; Shin, G.; Park, J.; Lim, C.J.; Chun, H.J.; Li, N.; Kim, D.H.; et al. The GIGANTEA-ENHANCED em LEVEL Complex Enhances Drought Tolerance via Regulation of Abscisic Acid Synthesis. *Plant Physiol.* **2020**, *184*, 443–458. [CrossRef]

80. Kim, W.Y.; Ali, Z.; Park, H.J.; Park, S.J.; Cha, J.Y.; Perez-Hormaeche, J.; Quintero, F.J.; Shin, G.; Kim, M.R.; Qiang, Z.; et al. Erratum: Corrigendum: Release of SOS2 kinase from sequestration with GIGANTEA determines salt tolerance in *Arabidopsis*. *Nat. Commun.* **2013**, *4*, 1820. [CrossRef]

81. Tseng, T.-S.; Salomé, P.A.; McClung, C.R.; Olszewski, N.E. SPINDLY and GIGANTEA Interact and Act in *Arabidopsis thaliana* Pathways Involved in Light Responses, Flowering, and Rhythms in Cotyledon Movements. *Plant Cell* **2004**, *16*, 1550–1563. [CrossRef]

82. Cha, J.Y.; Kim, J.; Kim, T.S.; Zeng, Q.; Wang, L.; Lee, S.Y.; Kim, W.Y.; Somers, D.E. GIGANTEA is a co-chaperone which facilitates maturation of ZEITLUPE in the *Arabidopsis* circadian clock. *Nat. Commun.* **2017**. [CrossRef] [PubMed]

83. Krahmer, J.; Goralogia, G.S.; Kubota, A.; Zardilis, A.; Johnson, R.S.; Song, Y.H.; MacCoss, M.J.; Le Bihan, T.; Halliday, K.J.; Imaizumi, T.; et al. Time-resolved interaction proteomics of the GIGANTEA protein under diurnal cycles in *Arabidopsis*. *FEBS Lett.* **2019**. [CrossRef] [PubMed]

84. Dröge-Laser, W.; Snoek, B.L.; Snel, B.; Weiste, C. The Arabidopsis bZIP transcription factor family—An update. *Curr. Opin. Plant Biol.* **2018**, *45*, 36–49. [CrossRef] [PubMed]

85. Abe, M.; Kosaka, S.; Shibuta, M.; Nagata, K.; Uemura, T.; Nakano, A.; Kaya, H. Transient activity of the florigen complex during the floral transition in *Arabidopsis thaliana*. *Development* **2019**, *146*. [CrossRef]

86. Wigge, P.A.; Kim, M.C.; Jaeger, K.E.; Busch, W.; Schmid, M.; Lohmann, J.U.; Weigel, D. Integration of spatial and temporal information during floral induction in *Arabidopsis*. *Science* **2005**, *309*, 1056–1059. [CrossRef]

87. Romera-Branchat, M.; Severing, E.; Pocard, C.; Ohr, H.; Vincent, C.; Née, G.; Martinez-Gallegos, R.; Jang, S.; Lalaguna, F.A.; Madrigal, P.; et al. Functional Divergence of the *Arabidopsis* Florigen-Interacting bZIP Transcription Factors FD and FDP. *Cell Rep.* **2020**, *31*. [CrossRef]

88. Zhang, B.; Li, C.; Li, Y.; Yu, H. Mobile TERMINAL FLOWER1 determines seed size in *Arabidopsis*. *Nat. Plants* **2020**, *6*, 1146–1157. [CrossRef]

89. Cheng, Z.J.; Zhao, X.Y.; Shao, X.X.; Wang, F.; Zhou, C.; Liu, Y.G.; Zhang, Y.; Zhang, X.S. Abscisic acid regulates early seed development in arabidopsis by ABI5-Mediated transcription of SHORT HYPOCOTYL UNDER BLUE1. *Plant Cell* **2014**, *26*, 1053–1068. [CrossRef]

90. Rodríguez-Martínez, J.A.; Reinke, A.W.; Bhimsaria, D.; Keating, A.E.; Ansari, A.Z. Combinatorial bZIP dimers display complex DNA-binding specificity landscapes. *Elife* **2017**, *6*. [CrossRef]

91. Cogato, A.; Meggio, F.; Migliorati, M.D.A.; Marinello, F. Extreme weather events in agriculture: A systematic review. *Sustainability* **2019**, *11*, 2547. [CrossRef]

92. Tardieu, F. Any trait or trait-related allele can confer drought tolerance: Just design the right drought scenario. *J. Exp. Bot.* **2012**, *63*, 25–31. [CrossRef] [PubMed]

93. Claeys, H.; Inzé, D. The agony of choice: How plants balance growth and survival under water-limiting conditions. *Plant Physiol.* **2013**, *162*, 1768–1779. [CrossRef] [PubMed]

94. Kenney, A.M.; Mckay, J.K.; Richards, J.H.; Juenger, T.E. Direct and indirect selection on flowering time, water-use efficiency (WUE, δ13C), and WUE plasticity to drought in *Arabidopsis thaliana*. *Ecol. Evol.* **2014**, *4*, 4505–4521. [CrossRef]

95. Franks, S.J. Plasticity and evolution in drought avoidance and escape in the annual plant *Brassica rapa*. *New Phytol.* **2011**, *190*, 249–257. [CrossRef] [PubMed]

96. Sherrard, M.E.; Maherali, H. The adaptive significance of drought escape in *Avena barbata*, an annual grass. *Evolution* **2006**, *60*, 2478. [CrossRef] [PubMed]

97. Shrestha, R.; Gómez-Ariza, J.; Brambilla, V.; Fornara, F. Molecular control of seasonal flowering in rice, arabidopsis and temperate cereals. *Ann. Bot.* **2014**, *114*, 1445–1458. [CrossRef]

98. Gómez-Ariza, J.; Galbiati, F.; Goretti, D.; Brambilla, V.; Shrestha, R.; Pappolla, A.; Courtois, B.; Fornara, F. Loss of floral repressor function adapts rice to higher latitudes in Europe. *J. Exp. Bot.* **2015**, *66*, 2027–2039. [CrossRef]

99. Du, H.; Huang, F.; Wu, N.; Li, X.; Hu, H.; Xiong, L. Integrative Regulation of Drought Escape through ABA-Dependent and -Independent Pathways in Rice. *Mol. Plant* **2018**, *11*, 584–597. [CrossRef]

100. Galbiati, F.; Chiozzotto, R.; Locatelli, F.; Spada, A.; Genga, A.; Fornara, F. Hd3a, RFT1 and Ehd1 integrate photoperiodic and drought stress signals to delay the floral transition in rice. *Plant. Cell Environ.* **2016**, *39*, 1982–1993. [CrossRef]

101. Wang, Y.; Lu, Y.; Guo, Z.; Ding, Y.; Ding, C. RICE CENTRORADIALIS 1, a TFL1-like Gene, Responses to Drought Stress and Regulates Rice Flowering Transition. *Rice* **2020**, *13*, 70. [CrossRef]

102. Miao, C.; Xiao, L.; Hua, K.; Zou, C.; Zhao, Y.; Bressan, R.A.; Zhu, J.K. Mutations in a subfamily of abscisic acid recepto genes promote rice growth and productivity. *Proc. Natl. Acad. Sci. USA* **2018**, *115*, 6058–6063. [CrossRef] [PubMed]

103. Kim, H.; Lee, K.; Hwang, H.; Bhatnagar, N.; Kim, D.Y.; Yoon, I.S.; Byun, M.O.; Kim, S.T.; Jung, K.H.; Kim, B.G. Overexpression of PYL5 in rice enhances drought tolerance, inhibits growth, and modulates gene expression. *J. Exp. Bot.* **2014**, *65*, 453–464. [CrossRef] [PubMed]

104. Xu, J.L.; Lafitte, H.R.; Gao, Y.M.; Fu, B.Y.; Torres, R.; Li, Z.K. QTLs for drought escape and tolerance identified in a set of random introgression lines of rice. *Theor. Appl. Genet.* **2005**. [CrossRef] [PubMed]

105. Guan, Y.S.; Serraj, R.; Liu, S.H.; Xu, J.L.; Ali, J.; Wang, W.S.; Venus, E.; Zhu, L.H.; Li, Z.K. Simultaneously improving yield under drought stress and non-stress conditions: A case study of rice (*Oryza sativa* L.). *J. Exp. Bot.* **2010**, *61*, 4145–4156. [CrossRef] [PubMed]

106. Nelson, D.E.; Repetti, P.P.; Adams, T.R.; Creelman, R.A.; Wu, J.; Warner, D.C.; Anstrom, D.C.; Bensen, R.J.; Castiglioni, P.P.; Donnarummo, M.G.; et al. Plant nuclear factor Y (NF-Y) B subunits confer drought tolerance and lead to improved corn yields on water-limited acres. *Proc. Natl. Acad. Sci. USA* **2007**, *104*, 16450–16455. [CrossRef] [PubMed]

107. Martignago, D.; Rico-Medina, A.; Blasco-Escámez, D.; Fontanet-Manzaneque, J.B.; Caño-Delgado, A.I. Drought Resistance by Engineering Plant Tissue-Specific Responses. *Front. Plant Sci.* **2020**, *10*, 1676. [CrossRef]

108. Shavrukov, Y.; Kurishbayev, A.; Jatayev, S.; Shvidchenko, V.; Zotova, L.; Koekemoer, F.; de Groot, S.; Soole, K.; Langridge, P. Early Flowering as a Drought Escape Mechanism in Plants: How Can It Aid Wheat Production? *Front. Plant Sci.* **2017**, *8*, 1950. [CrossRef]

Publisher's Note: MDPI stays neutral with regard to jurisdictional claims in published maps and institutional affiliations.

 International Journal of
Molecular Sciences

Review

Transcriptional Regulation of Protein Phosphatase 2C Genes to Modulate Abscisic Acid Signaling

Choonkyun Jung [1,2], Nguyen Hoai Nguyen [3] and Jong-Joo Cheong [4,*]

[1] Department of International Agricultural Technology and Crop Biotechnology, Institute/Green Bio Science and Technology, Seoul National University, Pyeongchang 25354, Korea; jasmin@snu.ac.kr

[2] Department of Plant Science, College of Agriculture and Life Sciences, Seoul National University, Seoul 08826, Korea

[3] Faculty of Biotechnology, Ho Chi Minh City Open University, Ho Chi Minh City 700000, Vietnam; nguyen.nhoai@ou.edu.vn

[4] Center for Food and Bioconvergence, Seoul National University, Seoul 08826, Korea

* Correspondence: cheongjj@snu.ac.kr; Tel.: +82-2-880-4888; Fax: +82-2-873-5260

Received: 27 October 2020; Accepted: 12 December 2020; Published: 14 December 2020

Abstract: The plant hormone abscisic acid (ABA) triggers cellular tolerance responses to osmotic stress caused by drought and salinity. ABA controls the turgor pressure of guard cells in the plant epidermis, leading to stomatal closure to minimize water loss. However, stomatal apertures open to uptake CO_2 for photosynthesis even under stress conditions. ABA modulates its signaling pathway via negative feedback regulation to maintain plant homeostasis. In the nuclei of guard cells, the clade A type 2C protein phosphatases (PP2Cs) counteract SnRK2 kinases by physical interaction, and thereby inhibit activation of the transcription factors that mediate ABA-responsive gene expression. Under osmotic stress conditions, PP2Cs bind to soluble ABA receptors to capture ABA and release active SnRK2s. Thus, PP2Cs function as a switch at the center of the ABA signaling network. ABA induces the expression of genes encoding repressors or activators of PP2C gene transcription. These regulators mediate the conversion of PP2C chromatins from a repressive to an active state for gene transcription. The stress-induced chromatin remodeling states of ABA-responsive genes could be memorized and transmitted to plant progeny; i.e., transgenerational epigenetic inheritance. This review focuses on the mechanism by which PP2C gene transcription modulates ABA signaling.

Keywords: abscisic acid; chromatin remodeling; drought; guard cell; osmotic stress; protein phosphatase 2C; salinity; stomata; stress memory; transgenerational inheritance

---

1. Introduction

The current global climate crisis has resulted in long spells of dry weather and a shortage of rainfall, and becomes a serious threat to crop productivity and food supply. Under drought conditions, the salt concentration increases as the moisture content decreases in the soil. Water deficit and salinity inflict osmotic stress on plant cells. Plants are not able to escape from adverse environments, and so respond to such stressful conditions by triggering physiological and cellular responses [1–3]. Most prominently, plants close stomatal apertures on the epidermis to limit transpiration and thereby prevent loss of water under osmotic stress conditions. A stomatal aperture is formed by two flanking guard cells that swell or deflate by regulating turgor pressure through ionic fluxes via ion channels anchored in the plasma membrane [4].

Under osmotic stress conditions, plants biosynthesize and accumulate abscisic acid (ABA), a sesquiterpenoid hormone [5]. Most importantly, ABA functions as a chemical messenger that induces numerous genes whose products are crucial for stomatal closure and the accumulation of osmoprotectants [6–8]. A previous transcriptomic study showed that more than half of the genes regulated by ABA treatment are also induced under drought or salinity conditions [9]. Likewise, ABA deficiency impairs osmotic stress regulation of gene expression [10]. Thus, it appears that osmotic stress-induced expression of the responsive genes is entirely dependent on ABA. Because plants encounter not only osmotic stress but also abnormal temperatures (heat and cold) and biotic stresses (pathogens and insects) in nature, ABA signaling is integrated with other ABA-independent signaling pathways [11,12].

ABA is mainly biosynthesized in vascular tissues and transported to sites of action, such as guard cells [13,14]. In guard cells, ABA molecules are perceived by receptors in the nucleus and cytosol, activating the sucrose non-fermenting 1-related protein kinase 2 (SnRK2) family of protein kinases [15,16]. In the nucleus, SnRK2s phosphorylate a number of transcription factors that activate transcription of the ABA-responsive genes whose products are implicated in stress responses and tolerance. Inversely, the clade A type 2C protein phosphatases (PP2Cs) counteract SnRK2s by physical interaction, exerting negative regulation of ABA signaling [17]. Under osmotic stress conditions, PP2Cs bind to ABA receptors to capture ABA, releasing and activating the SnRK2s. Thus, PP2Cs function as a switch at the center of the ABA signaling network.

In Arabidopsis, nine protein phosphatases are classified as clade A PP2Cs [18–20]. Six of them—ABA insensitive 1 (ABI1), ABI2, ABA hypersensitive germination 1 (AHG1), AHG3/PP2CA, hypersensitive to ABA1 (HAB1), and HAB2—are involved in ABA signaling in the osmotic stress response. The remaining three members, highly ABA-induced 1 (HAI1), PP2C1/HAI2, and HAI3, affected ABA-independent low water potential phenotypes, such as enhanced accumulation of osmoprotectants and suppression of the expression of abiotic stress-associated genes encoding dehydrins and late embryogenesis abundant proteins (LEAs) [21]. ABI1 and ABI2 are main components of ABA signaling under abiotic stresses and in developmental processes [22,23]. The dominant ABA response mutants of Arabidopsis, *abi1* and *abi2*, were originally isolated on the basis of their ABA insensitivity reflected in reduced seed dormancy and in symptoms of withering [24]. However, it was subsequently found that all of the knockout mutants of PP2C genes exhibited significant ABA hypersensitivity, indicating that they are negative regulators of ABA signaling. Recessive *hab1-1* mutants also showed enhanced ABA-responsive gene expression, increased ABA-mediated stomatal closure, and ABA-hypersensitivity in seed germination, indicating that HAB1 also negatively regulates ABA signaling [25,26].

ABA also plays pivotal roles in various physiological processes during the plant life cycle, including seed dormancy, germination, lateral root formation, light signaling convergence, and control of flowering time [5,7,12]. These functions of ABA are related to Ca^{2+} influx, the production of reactive oxygen species such as H_2O_2, ion transport, and electrical signaling [11,12,27]. During these processes, ABA signaling interacts antagonistically or synergistically with other hormonal signaling pathways mediated by auxin, cytokinin, ethylene, and jasmonates [7]. Thus, excess ABA impairs developmental processes such as senescence, as well as pollen fertility, and also leads to seed dormancy and susceptibility to diseases [28].

Stomata control transpiration and CO_2 uptake by optimizing the aperture size in response to various environmental and endogenous signals, including ABA, light, and CO_2 [29–32]. ABA causes stomatal closure, but light induces the opening of stomata to enhance CO_2 assimilation for photosynthesis. Plants often integrate osmotic stress and light signals simultaneously, and so the stomatal pores are opened and closed to maintain homeostasis.

Plants finely control the ABA concentration and ABA signaling during and after exposure to stressful conditions. The ABA levels in tissues are controlled by biosynthesis and catabolism [5]. In addition, the ABA signaling network can be desensitized by degradation of core proteins by the ubiquitin proteasome system [33]. In addition, plant cells modulate the ABA signaling pathway via PP2C-madiated negative feedback regulation.

ABA regulates the PP2C concentration by inducing the expression of genes encoding transcriptional repressors or activators. These transcriptional regulators compete with the PP2C gene promoters, inducing chromatin remodeling and thus the switch from a repressive to an active state. In this manner, ABA simultaneously activates positive and negative regulatory systems affecting its own signaling pathway. The chromatin state acquired for osmotic stress tolerance can be memorized and transmitted to newly developed cells during vegetative growth [34,35] and even inherited by the next generation of plants; i.e., transgenerational epigenetic inheritance [36,37].

In this article, we reviewed how plants modulate the ABA signaling pathway, focusing on the transcriptional regulation of PP2C gene expression by ABA. The biosynthesis, signaling mechanisms, and biological functions of ABA were recently reviewed comprehensively [38,39]. The epigenetic regulation of plant responses to abiotic stresses, including ABA treatment, drought, and salinity, were also reviewed in detail [40–42]. Kumar et al. [12] reviewed the integration of ABA signaling with other signaling pathways in development and plant stress responses.

2. Roles of PP2Cs in ABA Signaling

2.1. Negative Regulation of ABA Signaling

High levels of PP2Cs are part of the negative feedback mechanism that desensitizes plants to high ABA levels [43,44]. In the absence of ABA, PP2Cs physically interact with SnRK2s to form complexes (Figure 1A). In Arabidopsis, subgroup III SnRK2s are key regulators of ABA signaling [45,46]. There are 10 SnRK2 members in Arabidopsis; i.e., SnRK2.1–SnRK2.10. Among them, SnRK2.2, SnRK2.3, and SnRK2.6/OST1 are the strongest activators of ABA responses, and so are regarded as primary regulators of ABA signaling. The triple mutation (*snrk2.2/2.3/2.6*) largely blocked the major ABA responses [47]. ABI1 interacts with SnRK2.6/OST1, SnRK2.2, and SnRK2.3 in plants, resulting in the inactivation of downstream components; e.g., AREB/ABFs transcription factors and ion channels [46].

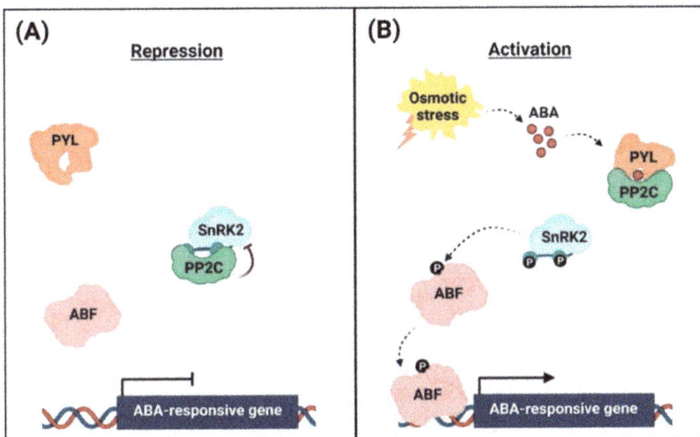

Figure 1. Abscisic acid (ABA) signaling pathway in the nuclei of guard cells. (**A**) Repression of ABA-responsive gene expression. In the absence, the clade A protein phosphatases (PP2Cs) physically interact with the sucrose non-fermenting 1-related protein kinases (SnRK2s) to reduce kinase activity via dephosphorylation. This inhibits the activity of ABRE-binding (AREB)/ABRE-binding factor (ABF) transcription factors and suppression of ABA-responsive gene transcription. (**B**) Activation of ABA-responsive gene expression. Under osmotic stress conditions, the interaction with ABA leads to conformational changes in the ABA receptors [PYR (pyrabactin resistance)/PYL (PYR-related)/RCAR (regulatory component of the ABA receptor)], allowing them to interact with PP2Cs. PP2Cs act as a coreceptor to capture ABA, thereby suppressing its phosphatase activity. This sequestrates PP2Cs from SnRK2s, and free SnRK2s phosphorylate the downstream transcription factors AREB/ABFs. The phosphorylated AREB/ABFs trigger the transcription of numerous ABA-responsive genes.

The SnRK2.6/OST1 was characterized as a critical limiting component in ABA regulation of stomatal apertures, ion channels, and NADPH oxidases in Arabidopsis guard cells [48]. PP2Cs dephosphorylate Ser175 in the activation loop of SnRK2.6, resulting in deactivation of the kinase [17]. Several PP2C-interacting factors, such as enhancer of ABA coreceptor 1 (EAR1) and PR5-like receptor kinase 2 (AtPR5K2), enhance the phosphatase activity of PP2Cs by phosphorylating them, and so modulate plant responses to drought stress [49,50].

2.2. Perception of ABA Signal

ABA molecules biosynthesized in vascular tissues are distantly transmitted to sites such as guard cells to activate the closure of stomata [13,14]. Multiple ABA transporters have been identified in Arabidopsis, including exporters (AtABCG25 and AtDTX50) and importers (AtABCG40 and AtAIT1) [51–55]. Guard cells themselves also biosynthesize ABA, which is sufficient for stomatal closure in response to low air humidity [56].

ABA molecules are perceived intracellularly by soluble receptors predominantly located in the nucleus and cytosol of guard cells [16,57]. A number of synonymous ABA receptors, e.g., pyrabactin resistance (PYR), PYR-related (PYL), and regulatory component of the ABA receptor (RCAR), have been identified as PP2C-interacting proteins in Arabidopsis [58–60]. PP2Cs have direct physical interactions with ABA and ABA receptors; these interactions are required for high-affinity binding of ABA [61,62]. Each PP2C functions as an ABA co-receptor within a holoreceptor complex that is constructed in combination with a particular PYR/PYL/RCAR.

The Arabidopsis genome contains 14 PYR/PYL/RCAR genes, which encode small proteins containing highly conserved amino acid residues [63]. All of them (except *PYL13*) are able to activate ABA-responsive gene expression. Transgenic lines expressing nuclear *PYR1* in an ABA-insensitive mutant background exhibited ABA responses, but cytosolic PYR1 was also required for full recovery of ABA responses [64]. PYL8/RCAR3 showed subcellular localization mainly in the cytosol and nucleus, and its overexpression led to enhanced drought resistance of Arabidopsis [65]. Guard cells express the six ABA receptor genes *PYR1*, *PYL1*, *PYL2*, *PYL4*, *PYL5*, and *PYL8* to mediate stomatal closure [66,67]. Arabidopsis mutants lacking three, four, five, and six of these PYR/PYL/RCAR genes (*pyr1/pyl1/pyl2/pyl4/pyl5/pyl8*) exhibited gradually increased stomatal conductance, indicating that this family of receptors quantitatively regulates the stomatal aperture [66]. Dittrich et al. [67] proposed that response specificity is achieved when the signals stimulate different members of the PYR/PYL/RCAR receptor family; PYL2 is sufficient for ABA-induced guard cell responses, whereas PYL4 and PYL5 are essential for the responses to CO_2. Different combinations of PYRs and PP2Cs influence ABA binding affinity, and therefore affect the ABA sensitivity of the whole plant [68,69].

ABA directly binds to the PYR/PYL/RCAR proteins [61,62,70,71]. ABA binding leads to conformational changes of the ABA receptors, which allows physical interaction with PP2Cs and inhibits phosphatase activity [72–74] (Figure 1B). Nishimura et al. [74] performed co-immunoprecipitation experiments in a transgenic Arabidopsis line stably transformed with yellow fluorescent protein (YFP)–*ABI1* fusion genes using a PYR1 antibody, and observed that the ABI1–PYR1 interaction was induced within 5 min after exogenous ABA application. Remarkable similarity was found in PP2C recognition between SnRK2 and ABA receptors [75,76]. In the absence of ABA, PP2C binds to the SnRK2 kinase domain and dephosphorylates Ser 175 in the activation loop. Upon perception of ABA, ABA receptor binds to PP2C by inserting the gate loop into the PP2C active cleft.

2.3. Regulation of ABA-Responsive Gene Expression

Upon the formation of PYL-ABA-PP2C complexes, SnRK2s dissociate from inactivated PP2Cs and recover their kinase activity. ABA treatment and osmotic stress stimulate phosphorylation of Ser 175 in the activation loop of SnRK2.6 [77]. When released from PP2C inhibition, SnRK2.6 autophosphorylates at Ser175 and Thr176 to recover full activity [76]. Free and active SnRK2s subsequently phosphorylate and activate downstream transcription factors in the nucleus and ion channels in the cytosol [57].

In the nucleus, the SnRK2-mediated phosphorylation of transcription factors results in the expression of numerous ABA-responsive genes. By analyzing the promoters of ABA-responsive genes, a conserved ABA-responsive element (ABRE; PyACGTGG/TC) was identified [78,79]. Subsequently, a number of ABRE-binding (AREB) proteins and ABRE-binding factors (ABFs) were identified by yeast one-hybrid screenings [80,81]. AREB/ABFs belong to the basic-domain leucine zipper (bZIP) transcription factor family and are colocalized with SnRK2s in plant cell nuclei [46]. Multiple conserved RxxS/T sites in AREB/ABFs are phosphorylated in an ABA-dependent manner [81–83].

Among the nine AREB/ABFs in Arabidopsis, ABF1, AREB1/ABF2, ABF3, and AREB2/ABF4 act as master transcription factors in ABA signaling for osmotic stress tolerance [84]. Overexpression of these genes in Arabidopsis resulted in ABA hypersensitivity and enhanced drought stress tolerance [85–87]. By contrast, the triple knockout mutant (*areb1/areb2/abf3*) displayed impaired expression of ABA- and osmotic stress-responsive genes, resulting in increased sensitivity to drought [88]. Fujii et al. [89] reconstituted ABA-triggered phosphorylation of ABF2/AREB1 in vitro by combining PYR1, ABI1, and SnRK2.6/OST1, demonstrating that PYR/PYL/RCAR receptors, PP2Cs, and SnRK2s constitute the core of the ABA signaling pathway.

3. Transcriptional Regulation of PP2C Gene Expression

3.1. ABA-Induced PP2C Gene Expression

ABA induces the expression of AREB/ABF genes, resulting in the accumulation of endogenous AREB/ABF proteins [6]. Concurrently, expression of the group-A PP2C genes is highly inducible in response to ABA and abiotic stresses [6,20]. The induction of PP2C gene expression may be an ABA desensitization mechanism modulating ABA signaling and maintaining plant homeostasis. Therefore, ABA upregulates genes encoding both positive and negative effectors of its signaling network.

The ABA-induced expression of PP2C genes is also mediated by AREB/ABFs. In response to salt stress, the transcript levels of PP2C genes (*ABI1*, *ABI2*, and *HAI1*) in an *abf3* mutant were markedly lower than those in wild-type plants [90], supporting a positive role for ABF3 in the activation of PP2C genes. A number of ABF3-binding sites, TCACGttt and ACACGgtt [91], are present in the promoter regions of these PP2C genes. In fact, a transcription factor hierarchy showed that ABF3 directly associates with the promoters of these genes [92]. Furthermore, Wang et al. [93] demonstrated that ABF transcription factors (i.e., ABF1 to ABF4) directly bind to the promoters of PP2C genes (*ABI1* and *ABI2*), and mediate rapid induction of their expression upon exogenous ABA treatment.

These data indicate that ABFs mediate ABA-induced expression of PP2C genes, thus playing a role in the negative feedback regulation of ABA signaling, in addition to the ABA-induced expression of ABA-responsive genes. Therefore, ABFs play dual in both the forward and backward regulation of ABA signaling. The ABF-mediated transcriptional upregulation of PP2Cs and PP2C-mediated inactivation of ABFs constitute a tight regulatory loop in ABA signaling modulation.

3.2. Repression of PP2C Gene Transcription

Under normal conditions, the expression of PP2C genes is maintained at basal levels, while under osmotically stressful conditions, the expression of PP2C genes is suppressed to enhance ABA signaling. A couple of MYB transcription factors were reported to act as repressors of PP2C gene transcription. For instance, *AtMYB44* transcripts accumulated under ABA treatment and abiotic stresses such as dehydration, low temperature, and salinity [94–96]. Microarray and northern blot analyses revealed that salt-induced expression of a group of PP2C genes, including *ABI1*, *ABI2*, *AtPP2CA*, *HAB1*, and *HAB2*, was significantly repressed in transgenic Arabidopsis overexpressing *AtMYB44* [94,95]. The transgenic plants showed increased sensitivity to ABA and more rapid ABA-induced stomatal closure. Under drought conditions, the transgenic Arabidopsis exhibited reduced rates of water loss and enhanced tolerance [94]. Furthermore, transgenic soybean [97] and rice seedlings [98] overexpressing *AtMYB44* exhibited significantly enhanced drought and salt stress tolerance. It appears that the enhanced osmotic stress tolerance of the transgenic plants was conferred by reduced expression of genes encoding PP2Cs that function as negative regulators of ABA-mediated stomatal closure. Cui et al. [99] also showed that the expression of a group of PP2C genes, such as *ABI1*, *ABI2*, and *PP2CA*, was suppressed in *AtMYB20*-overexpressing transgenic lines, but induced in *AtMYB20*-repression lines in response to salt treatment.

A number of the AtMYB44-binding sequences of AACnG [100] exist in transcription start site (TSS) regions of *ABI1*, *ABI2*, and *HAI1*. A chromatin immunoprecipitation (ChIP) assay demonstrated that AtMYB44 binds to the promoters of these genes under normal conditions to repress gene transcription [101]. In response to salt stress, AtMYB44 binding to PP2C promoters was significantly reduced, and the transcript levels of the genes were increased [90]. These results confirmed that AtMYB44 acts as a repressor of PP2C gene transcription. Such promoter-binding and repressive functions of AtMYB44 were also observed for *AtMYB44* [102] and *AtLEA4-5* [103].

A number of independent studies suggested that AtMYB44 physically interacts with ABA receptors. Jaradat et al. [96] observed that AtMYB44 (synonym MYBR1) physically interacts with PYL8 and represses ABA signaling in response to drought and senescence. Binding to PYL8 may block the interaction of AtMYB44 with PP2Cs or promoter of ABA-responsive genes. Li et al. [104] showed that AtMYB44 and ABI1 competed for binding to PYL9 and thereby reduced the inhibitory effect of the receptor on ABI1 phosphatase activity in the presence of ABA. These results suggest that AtMYB44 may act as a negative regulator of ABA signaling, which is inconsistent with its reported indirect positive role of suppressing PP2C gene transcription. Further studies are needed to explore the role of AtMYB44 as a positive or negative (or dual) regulator of ABA signaling.

4. Epigenetic Regulation of ABA Signaling

4.1. Epigenetic Regulation of ABA-Responsive Gene Transcription

In the chromatin of eukaryotic cells, genomic DNA is wrapped around a histone octamer consisting of H2A, H2B, H3, and H4 to form a nucleosome [105]. The access of RNA polymerase to the chromatin is regulated by competition between transcription factors and nucleosomes [106–108]. Thus, the chromatin around the gene transitions from a repressive state into an active state to enable access by RNA polymerase [109]. Chromatin remodeling is accompanied by histone modification (acetylation and methylation), DNA methylation, and microRNA generation, which take place mainly in the promoter region close to TSS [110,111]. Activators loaded on the promoter recruit co-activators and histone acetyltransferases (HATs) that acetylate the histones and relax DNA–histone binding in chromatin [112]. Inversely, repressors recruit corepressors associated with histone deacetylases (HDAs) so that nucleosomes bind tightly to DNA.

Epigenetic chromatin modification plays an important role in plant responses to osmotic stress [113–115]. Histone acetylation is involved in the transcriptional regulation of genes encoding PP2C family proteins, such as ABI1 and ABI2 [116]. Conversely, a histone deacetylation complex targets the promoters of the genes encoding PYL4, PYL5, and PYL6, thereby repressing gene expression [117]. Ryu et al. [118] reported that a histone deacetylation complex containing HDA19 binds to the promoter region of ABI3, and subsequently represses its expression. In addition, ABA enhances the methylation of promoter DNA, repressing the expression of ABA-repressive genes in Arabidopsis [119]. Moreover, ABA upregulates the expression of microRNAs in Arabidopsis, such as *miR159*, *miR393*, and *miR402* [120–122].

The switch/sucrose non-fermenting (SWI/SNF) chromatin remodeling complex regulates gene transcription in plants [123,124]. A subunit of the complex, BRAHMA (BRM), hydrolyzes ATP to supply the energy necessary to alter the interaction of nucleosomes with DNA, and thereby change the position and occupancy [125–127]. A whole-genome mapping and transcriptome analysis revealed that BRM complex occupies thousands of sites in the Arabidopsis genome, where it contributes to the activation or repression of gene transcription [128]. Han et al. [129] showed that the BRM complex in Arabidopsis represses ABA responses by affecting the stability of the associated nucleosome at a transcription factor (ABI5) locus, thus inactivating the gene. However, it is unclear how the BRM-containing SWI/SNF complexes access and occupy their target loci. Arabidopsis BRM contains several DNA-binding and nucleosome-binding regions, in addition to the AT-hook region [130]. In a study of vegetative development and flowering, BRM complex was recruited to specific loci by physical interaction with a plant-unique H3K27me3 demethylase that targets specific genomic loci [131].

Peirats-Llobet et al. [132] reported that SnRK2.2/2.3/2.6 kinases directly interacted with BRM, which led to phosphorylation and inhibition of its activity, while PP2CA-mediated dephosphorylation restored the ability of BRM to repress the ABA response. In this case, a phosphorylation-based switch mediated by SnRK2 and PP2C controls the BRM-associated chromatin remodeling state, thereby regulating the transcription of ABA-responsive genes.

4.2. Chromatin Remodeling for PP2C Gene Expression

AtMYB44 contains the amino acid sequence LxLxL, a putative ethylene-responsive element binding factor-associated amphiphilic repression (EAR) motif. Many studies have demonstrated physical interactions among EAR-containing repressors and TOPLESS (TPL) corepressor [133]. Ryu et al. [118] observed that the transcription factor BES1 forms a repressor complex with TPL and HDA19, directly facilitating the histone deacetylation of *ABI3* chromatin in Arabidopsis, although it remains unclear whether TPL–HDA19 interaction is direct or facilitated by adapter proteins. TPL-related (TPR) corepressors also recruit histone deacetylases such as HDA6 or HDA19, which are involved in various signaling pathways [134–136].

ChIP assay with transgenic Arabidopsis overexpressing the *AtMYB44-GFP* (green fluorescence protein) fusion gene revealed that AtMYB44 bound to PP2C gene (*ABI1*, *ABI2*, and *HAI1*) promoters to suppress gene transcription in a signal-independent manner [101]. Yeast two-hybrid and bimolecular fluorescence complementation (BiFC) assays demonstrated that AtMYB44 physically interacts with TPR1 and TPR3 corepressors through the EAR motif. Levels of histone H3 acetylation around the promoter and TSS proximal regions of *ABI1*, *ABI2*, and *HAI1* were markedly lower in *AtMYB44*-overexpressing transgenic plants than in wild-type plants. These results suggest that AtMYB44 forms a complex with TPR corepressors and recruits HDAs to suppress PP2C gene transcription (Figure 2A). Another repressor of PP2C gene transcription, AtMYB20 [99], also contains an EAR motif in the C-terminal side of the catalytic domain.

In response to salt stress, the AtMYB44 repressor was released and DNA–histone binding in nucleosomes were relaxed from the promoter regions [90], forcing chromatins to adopt an open structure (Figure 2B). Under these conditions, histone H3 acetylation (H3ac) around the TSS regions significantly increased. Wang et al. [93] demonstrated that ABFs bind to the promoters of PP2C genes and induce their transcription. Indeed, the salt-induced increases in PP2C gene (*ABI1*, *ABI2*, and *HAI1*) transcription were reduced in *abf3* plants [90]. In addition, whole Arabidopsis genome mapping revealed that BRM occupies, although does not directly bind to *ABI1* and *ABI2* gene promoters [128]. The Arabidopsis mutant *brm-3*, which shows moderately impaired BRM activity, produced more PP2C gene transcripts under salt stress conditions [90]. Thus, BRM contributes to the closed structure of PP2C chromatins, suppressing gene transcription.

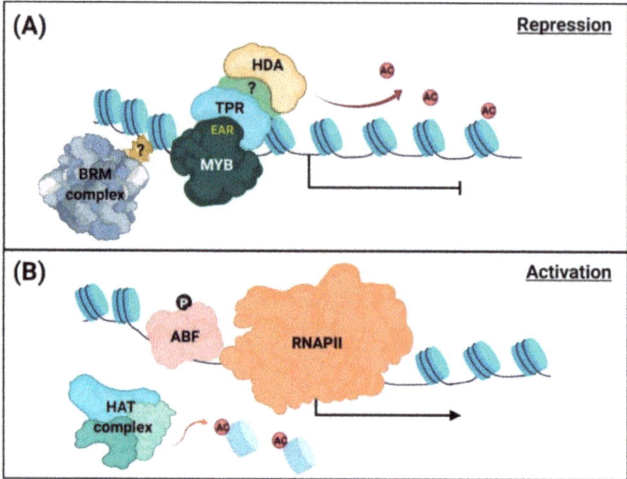

Figure 2. A working model of chromatin remodeling for regulation of PP2C gene transcription. (**A**) Repression of PP2C gene transcription. Under normal conditions, enhancer of ABA coreceptor (EAR) motif-containing MYB repressors (AtMYB44 and AtMYB20) interact with a TOPLESS-related corepressor (TPR), which recruits histone deacetylase (HDA) to suppress PP2C gene transcription. The chromatin remodeler, BRM-containing SWI/SNF complex, occupies the promoter and contributes to the repression of PP2C gene transcription. (**B**) Activation of PP2C gene transcription. Under osmotic stress conditions, the repressor is released from the promoter, and histone acetyltransferases (HATs) that acetylate the histones and relax DNA–histone binding in chromatin. Activator (AREB/ABFs) binds to the open promoter region, and RNA polymerase II (RNAPII) accesses and starts gene transcription.

4.3. Osmotic Stress Memory

A stressful condition enables plants to respond more promptly and strongly to repeated stress events [35,137]. For instance, Ding et al. [138] observed that Arabidopsis plants trained with previous dehydration events wilted much slower than non-trained plants under subsequent dehydration conditions. Virlouvet and Fromm [139] observed that the stomatal apertures in previously stressed Arabidopsis remain partially closed during a watered recovery period, facilitating reduced transpiration during subsequent dehydration stress. In addition, the rate-limiting ABA biosynthetic genes were expressed at much higher levels during watered recovery in the guard cells. Moreover, they performed a genetic analysis using mutants in the ABA signaling pathway, and found that SnRK2.2 and SnRK2.3 are important for stress memory of guard cells in the subsequent dehydration response.

In the memory responses, a subset of genes termed 'memory genes' are expressed at highly elevated or reduced levels during subsequent stress conditions. Numerous drought stress memory genes have been identified in Arabidopsis [140], maize [141,142], rice [143], potato [144], and soybean [145]. In Arabidopsis and soybean, drought-induced memory genes exhibiting elevated levels of transcripts include those involved in ABA-mediated tolerance responses to abiotic stresses, while the drought-repressed memory genes can be classified as light-harvesting- or photosynthesis-related genes [140,145]. When repeated dehydration stresses were imparted by air-drying, the Arabidopsis PP2C genes—including *ABI1*, *ABI2*, *HAB2*, *HAI2*, and *AtPP2CA*—did not exhibit expression patterns indicative of memory function [140]. By contrast, approximately 10 PP2C genes were identified as drought-induced memory genes in soybean grown in water-deprived soil [145]. The potential of stress memory to enhance crop productivity under drought conditions has been explored for a number of crops, including potato [146], wheat [147,148], and olive [149].

The most plausible mechanism underlying stress memory is changes in the chromatin architecture of memory gene loci [114,150,151]. For instance, histone methylation may act as a persistent epigenetic mark associated with transcriptional memory. H3K4me3 deposition in memory gene loci was higher than in non-memory genes after multiple exposures to drought stress [138,152]. Sani et al. [153] reported that hyperosmotic priming of Arabidopsis seedlings with transient mild salt treatment resulted in enhanced drought tolerance during a second stress exposure, leading to shortening and fractionation of H3K27me3 islands. Whatever the mechanism, such an epigenetically modified state may be transmitted mitotically to newly developed cells during the cell division process.

Furthermore, traits acquired under stressful conditions can be transmitted to progeny of the next generation [36,37,154]. Transgenerational epigenetic inheritance has been explored in crop breeding [155–157]. For instance, Raju et al. [158] developed an epigenetic breeding system in soybean for increased yield and stability, with RNAi suppression of a gene used to modulate developmental, defense, plant hormone, and abiotic stress response pathways. Verkest et al. [159] improved drought tolerance in canola by repeatedly selecting for increased drought tolerance in three generations. Tabassum et al. [160] observed that seed priming and transgenerational transmission improved tolerance to drought and salt stress in bread wheat. Zeng et al. [161] reported that multi-generation drought imposition mediated adaptation to drought condition in rice plants. Walter et al. [137] reported drought memory in grasses over an entire vegetation period, even after harvest and subsequent sprouting. However, net photosynthesis was reduced by 25% by recurrent drought treatment, which could have adverse effects on crop yield under more severe or longer droughts.

In general, the duration of a stress memory is relatively short, i.e., is limited to one generation [35,150,151]. Levels of the memory marker H3K4me3 in dehydration stress memory genes were elevated for 5 days [138]. This hampers application of stress memory to improve the stress tolerance of crops. In particular, although a number of studies have shown the involvement of epigenetic mechanisms, the principles underlying transgenerational inheritance are largely unknown [40,41]. Induced changes in the DNA methylation state were suggested as a possible mechanism by Zheng et al. [161], who observed that multi-generational drought stimulation induced the non-random appearance of epimutations and inheritance of high methylation state in advanced rice plant generations. As in animal cells, the acquired memory state could be reset (or forgot) during meiosis [162]. The mechanism by which plant cells overcome such resetting processes during meiosis and transmit the stress memory to progeny remains to be elucidated.

5. Conclusions and Perspectives

Drought and salinity are the most serious threats to crop productivity and food supply under global climate change. Therefore, understanding the mechanisms underlying osmotic stress tolerance and its application to crop breeding is an important topic in plant molecular science and biotechnology. ABA is a vital plant hormone that plays a key role in osmotic stress tolerance. ABA induces the closure of stomata in the epidermis, to limit transpiration and thereby prevent loss of water under osmotic stress conditions. The stomatal pores are open to uptake CO_2 for photosynthesis, and thereby maintain plant homeostasis. Therefore, it is not always favorable to artificially enhance ABA biosynthesis and signaling by gene modification or editing. It is essential to gain insight into the strategies that plants use in nature to deal with adverse environments without any negative effects on development or growth. In plant guard cells, PP2Cs counteract SnRKs for negative feedback regulation of ABA-induced stomatal closure. ABA induces both positive and negative mechanisms that modulate ABA responses by regulating PP2C gene transcription. Finally, plants encounter not only osmotic stress, but also temperature and biotic stresses. Therefore, communication between signaling pathways under different combinations of stresses should be more intensely investigated. Understanding the molecular mechanisms underlying stress memory and transgenerational inheritance might provide new methods to breed higher-quality crops that can withstand adverse climatic conditions.

Author Contributions: Writing—original draft preparation, C.J., N.H.N. and J.-J.C.; Writing—review and editing, J.-J.C. All authors have read and agreed to the published version of the manuscript.

Funding: This work was supported by the Creative-Pioneering Researchers Program of Seoul National University (to C.J.), the Next-Generation BioGreen 21 Program of the Rural Development Administration of Korea (grant number PJ013399 to C.J.), and the National Research Foundation of Korea (grant number 2020R111A1A01070089 to J.-J.C.).

Conflicts of Interest: The authors declare no conflict of interest.

Abbreviations

ABA	abscisic acid
ABF	ABRE-binding factor
AREB	ABRE-binding
BRM	BRAHMA
ChIP	chromatin immunoprecipitation
HDA	histone deacetyltransferase
EAR	ethylene-responsive element binding factor-associated amphiphilic repression
HAT	histone acetyltransferase
LEA	late embryogenesis abundant
PP2C	protein phosphatase 2C
PYL	PYR-related
PYR	pyrabactin resistance
RCAR	regulatory component of the ABA receptor
RNAPII	RNA polymerase II
SnRK2	sucrose non-fermenting 1-related protein kinase
SWI/SNF	switch/sucrose non-fermenting
TPL	TOPLESS
TPR	TPL-related
TSS	transcription start site

References

1. Gupta, A.; Rico-Medina, A.; Caño-Delgado, A.I. The physiology of plant responses to drought. *Science* **2020**, *368*, 266–269. [CrossRef]
2. Shinozaki, K.; Yamaguchi-Shinozaki, K. Molecular responses to drought and cold stress. *Curr. Opin. Biotechnol.* **1996**, *7*, 161–167. [CrossRef]
3. Shanker, A.K.; Maheswari, M.; Yadav, S.K.; Desai, S.; Bhanu, D.; Attal, N.B.; Venkateswarlu, B. Drought stress responses in crops. *Funct. Integr. Genom.* **2014**, *14*, 11–22. [CrossRef] [PubMed]
4. Lim, C.W.; Baek, W.; Jung, J.; Kim, J.-H.; Lee, S.C. Function of ABA in stomatal defense against biotic and drought stresses. *Int. J. Mol. Sci.* **2015**, *16*, 15251–15270. [CrossRef] [PubMed]
5. Nambara, E.; Marion-Poll, A. Abscisic acid biosynthesis and catabolism. *Annu. Rev. Plant Biol.* **2005**, *56*, 65–185. [CrossRef] [PubMed]
6. Fujita, Y.; Fujita, M.; Shinozaki, K.; Yamaguchi-Shinozaki, K. ABA-mediated transcriptional regulation in response to osmotic stress in plants. *J. Plant Res.* **2011**, *124*, 509–525. [CrossRef]
7. Sah, S.K.; Reddy, K.R.; Li, J. Abscisic acid and abiotic stress tolerance in crop plants. *Front. Plant Sci.* **2016**, *7*, 571. [CrossRef]
8. Vishwakarma, K.; Upadhyay, N.; Kumar, N.; Yadav, G.; Singh, J.; Mishra, R.K.; Kumar, V.; Verma, R.; Upadhyay, R.G.; Pandey, M.; et al. Abscisic acid signaling and abiotic stress tolerance in plants: A review on current knowledge and future prospects. *Front. Plant Sci.* **2017**, *8*, 161. [CrossRef]
9. Seki, M.; Ishida, J.; Narusaka, M.; Fujita, M.; Nanjo, T.; Umezawa, T.; Kamiya, A.; Nakajima, M.; Enju, A.; Sakurai, T.; et al. Monitoring the expression pattern of around 7,000 *Arabidopsis* genes under ABA treatments using a full-length cDNA microarray. *Funct. Integr. Genom.* **2002**, *2*, 282–291. [CrossRef]
10. Hoth, S.; Morgante, M.; Sanchez, J.-P.; Hanafey, M.K.; Tingey, S.V.; Chua, N.H. Genome-wide gene expression profiling in *Arabidopsis thaliana* reveals new targets of abscisic acid and largely impaired gene regulation in the *abi1-1* mutant. *J. Cell Sci.* **2002**, *115*, 4891–4900. [CrossRef]

11. Xiong, L.; Schumaker, K.S.; Zhu, J.-K. Cell signaling during cold, drought, and salt stress. *Plant Cell* **2002**, *14* (Suppl. 1), S165–S183. [CrossRef] [PubMed]

12. Kumar, M.; Kesawat, M.S.; Ali, A.; Lee, S.-C.; Gill, S.S.; Kim, H.U. Integration of abscisic acid signaling with other signaling pathways in plant stress responses and development. *Plants* **2019**, *8*, 592. [CrossRef] [PubMed]

13. Merilo, E.; Jalakas, P.; Laanemets, K.; Mohammadi, O.; Hõrak, H.; Kollist, H.; Brosché, M. Abscisic acid transport and homeostasis in the context of stomatal regulation. *Mol. Plant* **2015**, *8*, 1321–1333. [CrossRef] [PubMed]

14. Kuromori, T.; Seo, M.; Shinozaki, K. ABA transport and plant water stress responses. *Trends Plant Sci.* **2018**, *23*, 513–522. [CrossRef]

15. Umezawa, T.; Nakashima, K.; Miyakawa, T.; Kuromori, T.; Tanokura, M.; Shinozaki, K.; Yamaguchi-Shinozaki, K. Molecular basis of the core regulatory network in ABA responses: Sensing, signaling and transport. *Plant Cell Physiol.* **2010**, *51*, 1821–1839. [CrossRef]

16. Joshi-Saha, A.; Valon, C.; Leung, J. Abscisic acid signal off the STARTing block. *Mol. Plant* **2011**, *4*, 562–580. [CrossRef] [PubMed]

17. Umezawa, T.; Sugiyama, N.; Mizoguchi, M.; Hayashi, S.; Myouga, F.; Yamaguchi-Shinozaki, K.; Ishihama, Y.; Hirayama, T.; Shinozaki, K. Type 2C protein phosphatases directly regulate abscisic acid-activated protein kinases in *Arabidopsis*. *Proc. Natl. Acad. Sci. USA* **2009**, *106*, 17588–17593. [CrossRef] [PubMed]

18. Singh, A.; Pandey, A.; Srivastava, A.K.; Tran, L.-S.; Pandey, G.K. Plant protein phosphatases 2C: From genomic diversity to functional multiplicity and importance in stress management. *Crit. Rev. Biotechnol.* **2015**, *36*, 1023–1035. [CrossRef] [PubMed]

19. Schweighofer, A.; Hirt, H.; Meskiene, I. Plant PP2C phosphatases: Emerging functions in stress signaling. *Trends Plant Sci.* **2004**, *9*, 236–243. [CrossRef]

20. Xue, T.; Wang, D.; Zhang, S.; Ehlting, J.; Ni, F.; Jakab, S.; Zheng, C.; Zhong, Y. Genome-wide and expression analysis of protein phosphatase 2C in rice and *Arabidopsis*. *BMC Genom.* **2008**, *9*, 550. [CrossRef]

21. Bhaskara, G.B.; Nguyen, T.T.; Verslues, P.E. Unique drought resistance functions of the highly ABA-induced clade A protein phosphatase 2Cs. *Plant Physiol.* **2012**, *160*, 379–395. [CrossRef] [PubMed]

22. Gosti, F.; Beaudoin, N.; Serizet, C.; Webb, A.A.; Vartanian, N.; Giraudat, J. ABI1 protein phosphatase 2C is a negative regulator of abscisic acid signaling. *Plant Cell* **1999**, *11*, 1897–1909. [CrossRef] [PubMed]

23. Rodriguez, P.L.; Benning, G.; Grill, E. ABI2, a second protein phosphatase 2C involved in abscisic acid signal transduction in *Arabidopsis*. *FEBS Lett.* **1998**, *421*, 185–190. [CrossRef]

24. Koornneef, M.; Reuling, G.; Karssen, C.M. The isolation and characterization of abscisic acid-insensitive mutants of *Arabidopsis thaliana*. *Physiol. Plant.* **1984**, *61*, 377–383. [CrossRef]

25. Saez, A.; Apostolova, N.; Gonzalez-Guzman, M.; Gonzalez-Garcia, M.P.; Nicolas, C.; Lorenzo, O.; Rodriguez, P.L. Gain-of-function and loss-of-function phenotypes of the protein phosphatase 2C HAB1 reveal its role as a negative regulator of abscisic acid signalling. *Plant J.* **2004**, *37*, 354–369. [CrossRef] [PubMed]

26. Hirayama, T.; Shinozaki, K. Perception and transduction of abscisic acid signals: Keys to the function of the versatile plant hormone ABA. *Trends Plant Sci.* **2007**, *12*, 343–351. [CrossRef]

27. Mittler, R.; Blumwald, E. The role of ROS and ABA in systemic acquired acclimation. *Plant Cell* **2015**, *27*, 64–70. [CrossRef]

28. Gietler, M.; Fidler, J.; Labudda, M.; Nykiel, M. Abscisic acid—Enemy or savior in the response of cereals to abiotic and biotic stresses? *Int. J. Mol. Sci.* **2020**, *21*, 4607. [CrossRef]

29. Roelfsema, M.R.G.; Hedrich, R. In the light of stomatal opening: New insights into 'the Watergate'. *New Phytol.* **2005**, *167*, 665–691. [CrossRef]

30. Shimazaki, K.-I.; Doi, M.; Assmann, S.M.; Kinoshita, T. Light regulation of stomatal movement. *Annu. Rev. Plant Biol.* **2007**, *58*, 219–247. [CrossRef]

31. Kim, T.-H.; Böhmer, M.; Hu, H.; Nishimura, N.; Schroeder, J.I. Guard cell signal transduction network: Advances in understanding abscisic acid, CO_2, and Ca^{2+} signaling. *Annu. Rev. Plant Biol.* **2010**, *61*, 561–591. [CrossRef] [PubMed]

32. Negi, J.; Hashimoto-Sugimoto, M.; Kusumi, K.; Iba, K. New approaches to the biology of stomatal guard cells. *Plant Cell Physiol.* **2014**, *55*, 241–250. [CrossRef] [PubMed]

33. Ali, A.; Pardo, J.M.; Yun, D.-J. Desensitization of ABA-signaling: The swing from activation to degradation. *Front. Plant Sci.* **2020**, *11*, 379. [CrossRef] [PubMed]

34. Bruce, T.J.A.; Matthes, M.C.; Napier, J.A.; Pickett, J.A. Stressful "memories" of plants: Evidence and possible mechanisms. *Plant Sci.* **2007**, *173*, 603–608. [CrossRef]

35. Kinoshita, T.; Seki, M. Epigenetic memory for stress response and adaptation in plants. *Plant Cell Physiol.* **2014**, *55*, 1859–1863. [CrossRef] [PubMed]

36. Molinier, J.; Ries, G.; Zipfel, C.; Hohn, B. Transgeneration memory of stress in plants. *Nature* **2006**, *442*, 1046–1049. [CrossRef]

37. Quadrana, L.; Colot, V. Plant transgenerational epigenetics. *Annu. Rev. Genet.* **2016**, *50*, 467–491. [CrossRef]

38. Chen, K.; Li, G.J.; Bressan, R.A.; Song, C.P.; Zhu, J.K.; Zhao, Y. Abscisic acid dynamics, signaling, and functions in plants. *J. Integr. Plant Biol.* **2020**, *62*, 25–54. [CrossRef]

39. Sun, Y.; Pri-Tal, O.; Michaeli, D.; Mosquna, A. Evolution of abscisic acid signaling module and its perception. *Front. Plant Sci.* **2020**, *11*, 934. [CrossRef]

40. Ashapkin, V.V.; Kutueva, L.I.; Aleksandrushkina, N.I.; Vanyushin, B.F. Epigenetic mechanisms of plant adaptation to biotic and abiotic stresses. *Int. J. Mol. Sci.* **2020**, *21*, 7457. [CrossRef]

41. Chang, Y.-N.; Zhu, C.; Jiang, J.; Zhang, H.; Zhu, J.-K.; Duan, C.-G. Epigenetic regulation in plant abiotic stress responses. *J. Integr. Plant Biol.* **2020**, *62*, 563–580. [CrossRef] [PubMed]

42. Godwin, J.; Farrona, S. Plant epigenetic stress memory induced by drought: A physiological and molecular perspective. *Methods Mol. Biol.* **2020**, *2093*, 243–259. [CrossRef] [PubMed]

43. Rodriguez, P.L. Protein phosphatase 2C (PP2C) function in higher plants. *Plant Mol. Biol.* **1998**, *38*, 919–927. [CrossRef]

44. Fuchs, S.; Grill, E.; Meskiene, I.; Schweighofer, A. Type 2C protein phosphatases in plants. *FEBS J.* **2013**, *280*, 681–693. [CrossRef]

45. Fujii, H.; Zhu, J.-K. *Arabidopsis* mutant deficient in 3 abscisic acid-activated protein kinases reveals critical roles in growth, reproduction, and stress. *Proc. Natl. Acad. Sci. USA* **2009**, *106*, 8380–8385. [CrossRef]

46. Fujita, Y.; Nakashima, K.; Yoshida, T.; Katagiri, T.; Kidokoro, S.; Kanamori, N.; Umezawa, T.; Fujita, M.; Maruyama, K.; Ishiyama, K.; et al. Three SnRK2 protein kinases are the main positive regulators of abscisic acid signaling in response to water stress in *Arabidopsis*. *Plant Cell Physiol.* **2009**, *50*, 2123–2132. [CrossRef]

47. Nakashima, K.; Fujita, Y.; Kanamori, N.; Katagiri, T.; Umezawa, T.; Kidokoro, S.; Maruyama, K.; Yoshida, T.; Ishiyama, K.; Kobayashi, M.; et al. Three *Arabidopsis* SnRK2 protein kinases, SRK2D/SnRK2.2, SRK2E/SnRK2.6/OST1 and SRK2I/SnRK2.3, involved in ABA signaling are essential for the control of seed development and dormancy. *Plant Cell Physiol.* **2009**, *50*, 1345–1363. [CrossRef]

48. Acharya, B.R.; Jeon, B.W.; Zhang, W.; Assmann, S.M. Open Stomata 1 (OST1) is limiting in abscisic acid responses of *Arabidopsis* guard cells. *New Phytol.* **2013**, *200*, 1049–1063. [CrossRef]

49. Wang, K.; He, J.; Zhao, Y.; Wu, T.; Zhou, X.; Ding, Y.; Kong, L.; Wang, X.; Wang, Y.; Li, J.; et al. EAR1 negatively regulates ABA signaling by enhancing 2C protein phosphatase activity. *Plant Cell* **2018**, *30*, 815–834. [CrossRef]

50. Baek, D.; Kim, M.C.; Kumar, D.; Park, B.; Cheong, M.S.; Choi, W.; Park, H.C.; Chun, H.J.; Park, H.J.; Lee, S.Y.; et al. AtPR5K2, a PR5-Like receptor kinase, modulates plant responses to drought stress by phosphorylating protein phosphatase 2Cs. *Front. Plant Sci.* **2019**, *10*, 1146. [CrossRef]

51. Kuromori, T.; Miyaji, T.; Yabuuchi, H.; Shimizu, H.; Sugimoto, E.; Kamiya, A.; Moriyama, Y.; Shinozaki, K. ABC transporter AtABCG25 is involved in abscisic acid transport and responses. *Proc. Natl. Acad. Sci. USA* **2010**, *107*, 2361–2366. [CrossRef]

52. Kang, J.; Hwang, J.U.; Lee, M.; Kim, Y.Y.; Assmann, S.M.; Martinoia, E.; Lee, Y. PDR-type ABC transporter mediates cellular uptake of the phytohormone abscisic acid. *Proc. Natl. Acad. Sci. USA* **2010**, *107*, 2355–2360. [CrossRef] [PubMed]

53. Kanno, Y.; Hanada, A.; Chiba, Y.; Ichikawa, T.; Nakazawa, M.; Matsui, M.; Koshiba, T.; Kamiya, Y.; Seo, M. Identification of an abscisic acid transporter by functional screening using the receptor complex as a sensor. *Proc. Natl. Acad. Sci. USA* **2012**, *109*, 9653–9658. [CrossRef] [PubMed]

54. Ng, L.M.; Melcher, K.; Teh, B.T.; Xu, H.E. Abscisic acid perception and signaling: Structural mechanisms and applications. *Acta Pharmacol. Sin.* **2014**, *35*, 567–584. [CrossRef] [PubMed]

55. Zhang, H.; Zhu, H.; Pan, Y.; Yu, Y.; Luan, S.; Li, L. A DTX/MATE-type transporter facilitates abscisic acid efflux and modulates ABA sensitivity and drought tolerance in *Arabidopsis*. *Mol. Plant* **2014**, *7*, 1522–1532. [CrossRef] [PubMed]

56. Bauer, H.; Ache, P.; Lautner, S.; Fromm, J.; Hartung, W.; Al-Rasheid, K.A.S.; Sonnewald, S.; Sonnewald, U.; Kneitz, S.; Lachmann, N.; et al. The stomatal response to reduced relative humidity requires guard cell-autonomous ABA synthesis. *Curr. Biol.* **2013**, *23*, 53–57. [CrossRef]

57. Dong, T.; Park, Y.; Hwang, I. Abscisic acid: Biosynthesis, inactivation, homoeostasis and signaling. *Essays Biochem.* **2015**, *58*, 29–48. [CrossRef]

58. Cutler, S.R.; Rodriguez, P.L.; Finkelstein, R.R.; Abrams, S.R. Abscisic acid: Emergence of a core signaling network. *Annu. Rev. Plant Biol.* **2010**, *61*, 651–679. [CrossRef]

59. Melcher, K.; Zhou, X.E.; Xu, H.E. Thirsty plants and beyond: Structural mechanisms of abscisic acid perception and signaling. *Curr. Opin. Struct. Biol.* **2010**, *20*, 722–729. [CrossRef]

60. Guo, J.; Yang, X.; Weston, D.J.; Chen, J.-G. Abscisic acid receptors: Past, present and future. *J. Integr. Plant Biol.* **2011**, *53*, 469–479. [CrossRef]

61. Miyazono, K.-I.; Miyakawa, T.; Sawano, Y.; Kubota, K.; Kang, H.-J.; Asano, A.; Miyauchi, Y.; Takahashi, M.; Zhi, Y.; Fujita, Y.; et al. Structural basis of abscisic acid signaling. *Nature* **2009**, *462*, 609–614. [CrossRef]

62. Yin, P.; Fan, H.; Hao, Q.; Yuan, X.; Wu, D.; Pang, Y.; Yan, C.; Li, W.; Wang, J.; Yan, N. Structural insights into the mechanism of abscisic acid signaling by PYL proteins. *Nat. Struct. Mol. Biol.* **2009**, *16*, 1230–1236. [CrossRef]

63. Klingler, J.P.; Batelli, G.; Zhu, J.-K. ABA receptors: The START of a new paradigm in phytohormone signalling. *J. Exp. Bot.* **2010**, *61*, 3199–3210. [CrossRef]

64. Park, E.; Kim, T.-H. Production of ABA responses requires both the nuclear and cytoplasmic functional involvement of PYR1. *Biochem. Biophys. Res. Commun.* **2017**, *484*, 34–39. [CrossRef] [PubMed]

65. Saavedra, X.; Modrego, A.; Rodríguez, D.; González-García, M.P.; Sanz, L.; Nicolás, G.; Lorenzo, O. The nuclear interactor PYL8/RCAR3 of *Fagus sylvatica* FsPP2C1 is a positive regulator of abscisic acid signaling in seeds and stress. *Plant Physiol.* **2010**, *152*, 133–150. [CrossRef]

66. Gonzalez-Guzman, M.; Pizzio, G.A.; Antoni, R.; Vera-Sirera, F.; Merilo, E.; Bassel, G.W.; Fernández, M.A.; Holdsworth, M.J.; Perez-Amador, M.A.; Kollist, H.; et al. *Arabidopsis* PYR/PYL/RCAR receptors play a major role in quantitative regulation of stomatal aperture and transcriptional response to abscisic acid. *Plant Cell* **2012**, *24*, 2483–2496. [CrossRef]

67. Dittrich, M.; Mueller, H.M.; Bauer, H.; Peirats-Llobet, M.; Rodriguez, P.L.; Geilfus, C.M.; Carpentier, S.C.; Al Rasheid, K.A.S.; Kollist, H.; Merilo, E.; et al. The role of *Arabidopsis* ABA receptors from the PYR/PYL/RCAR family in stomatal acclimation and closure signal integration. *Nat. Plants* **2019**, *5*, 1002–1011. [CrossRef]

68. Szostkiewicz, I.; Richter, K.; Kepka, M.; Demmel, S.; Ma, Y.; Korte, A.; Assaad, F.F.; Christmann, A.; Grill1, E. Closely related receptor complexes differ in their ABA selectivity and sensitivity. *Plant J.* **2010**, *61*, 25–35. [CrossRef]

69. Tischer, S.V.; Wunschel, C.; Papacek, M.; Kleigrewe, K.; Hofmann, T.; Christmann, A.; Grill, E. Combinatorial interaction network of abscisic acid receptors and coreceptors from *Arabidopsis thaliana*. *Proc. Natl. Acad. Sci. USA* **2017**, *114*, 10280–10285. [CrossRef] [PubMed]

70. Santiago, J.; Dupeux, F.; Round, A.; Antoni, R.; Park, S.Y.; Jamin, M.; Cutler, S.R.; Rodriguez, P.L.; Marquez, J.A. The abscisic acid receptor PYR1 in complex with abscisic acid. *Nature* **2009**, *462*, 665–668. [CrossRef] [PubMed]

71. Zhang, X.L.; Jiang, L.; Xin, Q.; Liu, Y.; Tan, J.X.; Chen, Z.Z. Structural basis and functions of abscisic acid receptors PYLs. *Front Plant Sci.* **2015**, *6*, 88. [CrossRef] [PubMed]

72. Ma, Y.; Szostkiewicz, I.; Korte, A.; Moes, D.; Yang, Y.; Christmann, A.; Grill, E. Regulators of PP2C phosphatase activity function as abscisic acid sensors. *Science* **2009**, *324*, 1064–1068. [CrossRef] [PubMed]

73. Park, S.Y.; Fung, P.; Nishimura, N.; Jensen, D.R.; Fujii, H.; Zhao, Y.; Lumba, S.; Santiago, J.; Rodrigues, A.; Chow, T.F.; et al. Abscisic acid inhibits type 2C protein phosphatases via the PYR/PYL family of START proteins. *Science* **2009**, *324*, 1068–1071. [CrossRef] [PubMed]

74. Nishimura, N.; Sarkeshik, A.; Nito, K.; Park, S.-Y.; Wang, A.; Carvalho, P.C.; Lee, S.; Caddell, D.F.; Cutler, S.R.; Chory, J.; et al. PYR/PYL/RCAR family members are major in-vivo ABI1 protein phosphatase 2C-interacting proteins in *Arabidopsis*. *Plant J.* **2010**, *61*, 290–299. [CrossRef]

75. Soon, F.-F.; Ng, L.-M.; Zhou, X.E.; West, G.M.; Kovach, A.; Tan, M.H.E.; Suino-Powell, K.M.; He, Y.; Xu, Y.; Chalmers, M.J.; et al. Molecular mimicry regulates ABA signaling by SnRK2 kinases and PP2C phosphatases. *Science* **2012**, *335*, 85–88. [CrossRef]

76. Fujii, H.; Zhu, J.-K. Osmotic stress signaling via protein kinases. *Cell. Mol. Life Sci.* **2012**, *69*, 3165–3173. [CrossRef]

77. Belin, C.; De Franco, P.-O.; Bourbousse, C.; Chaignepain, S.; Schmitter, J.M.; Vavasseur, A.; Giraudat, J.; Barbier-Brygoo, H.; Thomine, S. Identification of features regulating OST1 kinase activity and OST1 function in guard cells. *Plant Physiol.* **2006**, *141*, 1316–1327. [CrossRef]

78. Giraudat, J.; Parcy, F.; Bertauche, N.; Gosti, F.; Leung, J. Current advances in abscisic acid action and signaling. *Plant Mol. Biol.* **1994**, *26*, 1557–1577. [CrossRef]

79. Busk, P.K.; Pagès, M. Regulation of abscisic acid-induced transcription. *Plant Mol. Biol.* **1998**, *37*, 425–435. [CrossRef]

80. Choi, H.; Hong, J.; Ha, J.; Kang, J.; Kim, S.Y. ABFs, a family of ABA-responsive element binding factors. *J. Biol. Chem.* **2000**, *275*, 1723–1730. [CrossRef]

81. Uno, Y.; Furihata, T.; Abe, H.; Yoshida, R.; Shinozaki, K.; Yamaguchi-Shinozaki, K. *Arabidopsis* basic leucine zipper transcription factors involved in an abscisic acid-dependent signal transduction pathway under drought and high-salinity conditions. *Proc. Natl Acad. Sci. USA* **2000**, *97*, 11632–11637. [CrossRef] [PubMed]

82. Furihata, T.; Maruyama, K.; Fujita, Y.; Umezawa, T.; Yoshida, R.; Shinozaki, K.; Yamaguchi-Shinozaki, K. Abscisic acid-dependent multisite phosphorylation regulates the activity of a transcription activator AREB1. *Proc. Natl. Acad. Sci. USA* **2006**, *103*, 1988–1993. [CrossRef] [PubMed]

83. Fujii, H.; Verslues, P.E.; Zhu, J.K. Identification of two protein kinases required for abscisic acid regulation of seed germination, root growth, and gene expression in *Arabidopsis*. *Plant Cell* **2007**, *19*, 485–494. [CrossRef]

84. Yoshida, T.; Fujita, Y.; Maruyama, K.; Mogami, J.; Todaka, D.; Shinozaki, K.; Yamaguchi-Shinozaki, K. Four *Arabidopsis* AREB/ABF transcription factors function predominantly in gene expression downstream of SnRK2 kinases in abscisic acid signalling in response to osmotic stress. *Plant Cell Environ.* **2015**, *38*, 35–49. [CrossRef] [PubMed]

85. Kang, J.Y.; Choi, H.I.; Im, M.Y.; Kim, S.Y. *Arabidopsis* basic leucine zipper proteins that mediate stress-responsive abscisic acid signaling. *Plant Cell* **2002**, *14*, 343–357. [CrossRef] [PubMed]

86. Kim, S.; Kang, J.Y.; Cho, D.I.; Park, J.H.; Kim, S.Y. ABF2, an ABRE-binding bZIP factor, is an essential component of glucose signaling and its overexpression affects multiple stress tolerance. *Plant J.* **2004**, *40*, 75–87. [CrossRef]

87. Fujita, Y.; Fujita, M.; Satoh, R.; Maruyama, K.; Parvez, M.M.; Seki, M.; Hiratsu, K.; Ohme-Takagi, M.; Shinozaki, K.; Yamaguchi-Shinozaki, K. AREB1 is a transcription activator of novel ABRE-dependent ABA signaling that enhances drought stress tolerance in *Arabidopsis*. *Plant Cell* **2005**, *17*, 3470–3488. [CrossRef]

88. Yoshida, T.; Fujita, Y.; Sayama, H.; Kidokoro, S.; Maruyama, K.; Mizoi, J.; Shinozaki, K.; Yamaguchi-Shinozaki, K. AREB1, AREB2, and ABF3 are master transcription factors that cooperatively regulate ABRE-dependent ABA signaling involved in drought stress tolerance and require ABA for full activation. *Plant J.* **2010**, *61*, 672–685. [CrossRef]

89. Fujii, H.; Chinnusamy, V.; Rodrigues, A.; Rubio, S.; Antoni, R.; Park, S.-Y.; Cutler, S.R.; Sheen, J.; Rodriguez, P.L.; Zhu, J.-K. In vitro reconstitution of an ABA signaling pathway. *Nature* **2009**, *462*, 660–664. [CrossRef]

90. Nguyen, N.H.; Jung, C.; Cheong, J.-J. Chromatin remodeling for the transcription of type 2C protein phosphatase genes in response to salt stress. *Plant Physiol. Biochem.* **2019**, *141*, 325–331. [CrossRef]

91. Weirauch, M.T.; Yang, A.; Albu, M.; Cote, A.G.; Montenegro-Montero, A.; Drewe, P.; Najafabadi, H.S.; Lambert, S.A.; Mann, I.; Cook, K.; et al. Determination and inference of eukaryotic transcription factor sequence specificity. *Cell* **2014**, *158*, 1431–1443. [CrossRef] [PubMed]

92. Song, L.; Huang, S.C.; Wise, A.; Castanon, R.; Nery, J.R.; Chen, H.; Watanabe, M.; Thomas, J.; Bar-Joseph, Z.; Ecker, J.R. A transcription factor hierarchy defines an environmental stress response network. *Science* **2016**, *354*, aag1550. [CrossRef] [PubMed]

93. Wang, X.; Guo, C.; Peng, J.; Li, C.; Wan, F.; Zhang, S.; Zhou, Y.; Yan, Y.; Qi, L.; Sun, K.; et al. ABRE-BINDING FACTORS play a role in the feedback regulation of ABA signaling by mediating rapid ABA induction of ABA co-receptor genes. *New Phytol.* **2019**, *221*, 341–355. [CrossRef] [PubMed]

94. Jung, C.; Seo, J.S.; Han, S.W.; Koo, Y.J.; Kim, C.H.; Song, S.I.; Nahm, B.H.; Choi, Y.D.; Cheong, J.-J. Overexpression of *AtMYB44* enhances stomatal closure to confer abiotic stress tolerance in transgenic *Arabidopsis*. *Plant Physiol.* **2008**, *146*, 623–635. [CrossRef] [PubMed]

95. Jung, C.; Shim, J.S.; Seo, J.S.; Lee, H.Y.; Kim, C.H.; Choi, Y.D.; Cheong, J.-J. Non-specific phytohormonal induction of AtMYB44 and suppression of jasmonate-responsive gene activation in *Arabidopsis thaliana*. *Mol. Cells* **2010**, *29*, 71–76. [CrossRef]

96. Jaradat, M.R.; Feurtado, J.A.; Huang, D.; Lu, Y.; Cutler, A.J. Multiple roles of the transcription factor AtMYBR1/AtMYB44 in ABA signaling, stress responses, and leaf senescence. *BMC Plant Biol.* **2013**, *13*, 192. [CrossRef] [PubMed]

97. Seo, J.S.; Sohn, H.B.; Noh, K.; Jung, C.; An, J.H.; Donovan, C.M.; Somers, D.A.; Kim, D.I.; Jeong, S.-C.; Kim, C.-G.; et al. Expression of the *Arabidopsis AtMYB44* gene confers drought/salt-stress tolerance in transgenic soybean. *Mol. Breed.* **2012**, *29*, 601–608. [CrossRef]

98. Joo, J.; Oh, N.-I.; Nguyen, N.H.; Lee, Y.H.; Kim, Y.-K.; Song, S.I.; Cheong, J.-J. Intergenic transformation of *AtMYB44* confers drought stress tolerance in rice seedlings. *Appl. Biol. Chem.* **2017**, *60*, 447–455. [CrossRef]

99. Cui, M.H.; Yoo, K.S.; Hyoung, S.; Nguyen, H.T.K.; Kim, Y.Y.; Kim, H.J.; Ok, S.H.; Yoo, S.D.; Shin, J.S. An *Arabidopsis* R2R3-MYB transcription factor, AtMYB20, negatively regulates type 2C serine/threonine protein phosphatases to enhance salt tolerance. *FEBS Lett.* **2013**, *587*, 1773–1778. [CrossRef]

100. Jung, C.; Kim, Y.-K.; Oh, N.I.; Shim, J.S.; Seo, J.S.; Choi, Y.D.; Nahm, B.H.; Cheong, J.-J. Quadruple 9-mer-based protein binding microarray analysis confirms AACnG as the consensus nucleotide sequence sufficient for the specific binding of AtMYB44. *Mol. Cells* **2012**, *34*, 531–537. [CrossRef]

101. Nguyen, N.H.; Cheong, J.-J. AtMYB44 interacts with TOPLESS-RELATED corepressors to suppress *protein phosphatase 2C* gene transcription. *Biochem. Biophys. Res. Commun.* **2018**, *507*, 437–442. [CrossRef] [PubMed]

102. Nguyen, N.H.; Cheong, J.-J. H2A.Z-containing nucleosomes are evicted to activate *AtMYB44* transcription in response to salt stress. *Biochem. Biophys. Res. Commun.* **2018**, *499*, 1039–1043. [CrossRef] [PubMed]

103. Nguyen, N.H.; Nguyen, C.T.T.; Jung, C.; Cheong, J.-J. AtMYB44 suppresses transcription of the late embryogenesis abundant protein gene *AtLEA4-5*. *Biochem. Biophys. Res. Commun.* **2019**, *511*, 931–934. [CrossRef] [PubMed]

104. Li, D.; Li, Y.; Zhang, L.; Wang, X.; Zhao, Z.; Tao, Z.; Wang, J.; Wang, J.; Lin, M.; Li, X.; et al. *Arabidopsis* ABA receptor RCAR1/PYL9 interacts with an R2R3-type MYB transcription factor, AtMYB44. *Int. J. Mol. Sci.* **2014**, *15*, 8473–8490. [CrossRef] [PubMed]

105. Zhang, T.; Zhang, W.; Jiang, J. Genome-wide nucleosome occupancy and positioning and their impact on gene expression and evolution in plants. *Plant Physiol.* **2015**, *168*, 1406–1416. [CrossRef]

106. Struhl, K.; Segal, E. Determinants of nucleosome positioning. *Nat. Struct. Mol. Biol.* **2013**, *20*, 267–273. [CrossRef]

107. Venkatesh, S.; Workman, J.L. Histone exchange, chromatin structure and the regulation of transcription. *Nat. Rev. Mol. Cell Biol.* **2015**, *16*, 178–189. [CrossRef]

108. Klemm, S.L.; Shipony, Z.; Greenleaf, W.J. Chromatin accessibility and the regulatory epigenome. *Nat. Rev. Genet.* **2019**, *20*, 207–220. [CrossRef]

109. Cairns, B.R. The logic of chromatin architecture and remodeling at promoters. *Nature* **2009**, *461*, 193–198. [CrossRef]

110. Yamamuro, C.; Zhu, J.K.; Yang, Z. Epigenetic modifications and plant hormone action. *Mol. Plant* **2016**, *9*, 57–70. [CrossRef]

111. Bartel, D.P. MicroRNA target recognition and regulatory functions. *Cell* **2009**, *136*, 215–233. [CrossRef] [PubMed]

112. Davie, J.R. Covalent modifications of histones: Expression from chromatin templates. *Curr. Opin. Genet. Dev.* **1998**, *8*, 173–178. [CrossRef]

113. To, T.K.; Kim, J.M. Epigenetic regulation of gene responsiveness in *Arabidopsis*. *Front. Plant Sci.* **2014**, *4*, 548. [CrossRef] [PubMed]

114. Kim, J.-M.; Sasaki, T.; Ueda, M.; Sako, K.; Seki, M. Chromatin changes in response to drought, salinity, heat, and cold stresses in plants. *Front. Plant Sci.* **2015**, *8*, 114. [CrossRef]

115. Ueda, M.; Seki, M. Histone modifications form epigenetic regulatory networks to regulate abiotic stress response. *Plant Physiol.* **2020**, *182*, 15–26. [CrossRef]

116. Luo, M.; Liu, X.; Singh, P.; Cui, Y.; Zimmerli, L.; Wu, K. Chromatin modifications and remodeling in plant abiotic stress responses. *Biochim. Biophys. Acta* **2012**, *1819*, 129–136. [CrossRef]

117. Mehdi, S.; Derkacheva, M.; Ramström, M.; Kralemann, L.; Bergquist, J.; Hennig, L. The WD40 domain protein MSI1 functions in a histone deacetylase complex to fine-tune abscisic acid signaling. *Plant Cell* **2016**, *28*, 42–54. [CrossRef]

118. Ryu, H.; Cho, H.; Bae, W.; Hwang, I. Control of early seedling development by BES1/TPL/HDA19-mediated epigenetic regulation of ABI3. *Nat. Commun.* **2014**, *5*, 4138. [CrossRef]

119. Gohlke, J.; Scholz, C.-J.; Kneitz, S.; Weber, D.; Fuchs, J.; Hedrich, R.; Deeken, R. DNA methylation mediated control of gene expression is critical for development of crown gall tumors. *PLoS Genet.* **2013**, *9*, e1003267. [CrossRef] [PubMed]

120. Sunkar, R.; Zhu, J.-K. Novel and stress-regulated microRNAs and other small RNAs from *Arabidopsis*. *Plant Cell* **2004**, *16*, 2001–2019. [CrossRef]

121. Reyes, J.L.; Chua, N.H. ABA induction of miR159 controls transcript levels of two MYB factors during *Arabidopsis* seed germination. *Plant J.* **2007**, *49*, 592–606. [CrossRef] [PubMed]

122. Li, W.X.; Oono, Y.; Zhu, J.; He, X.J.; Wu, J.M.; Iida, K.; Lu, X.-Y.; Cui, X.; Jin, H.; Zhu, J.-K. The *Arabidopsis* NFYA5 transcription factor is regulated transcriptionally and posttranscriptionally to promote drought resistance. *Plant Cell* **2008**, *20*, 2238–2251. [CrossRef] [PubMed]

123. Han, S.-K.; Wu, M.-F.; Cui, S.; Wagner, D. Roles and activities of chromatin remodeling ATPases in plants. *Plant J.* **2015**, *83*, 62–77. [CrossRef] [PubMed]

124. Thouly, C.; Masson, M.L.; Lai, X.; Carles, C.C.; Vachon, G. Unwinding BRAHMA functions in plants. *Genes* **2020**, *11*, 90. [CrossRef]

125. Clapier, C.R.; Cairns, B.R. The biology of chromatin remodeling complexes. *Annu. Rev. Biochem.* **2009**, *78*, 273–304. [CrossRef]

126. Hargreaves, D.C.; Crabtree, G.R. ATP-dependent chromatin remodeling: Genetics, genomics and mechanisms. *Cell Res.* **2011**, *21*, 396–420. [CrossRef]

127. Narlikar, G.J.; Sundaramoorthy, R.; Owen-Hughes, T. Mechanisms and functions of ATP-dependent chromatin-remodeling enzymes. *Cell* **2013**, *154*, 490–503. [CrossRef]

128. Archacki, R.; Yatusevich, R.; Buszewicz, D.; Krzyczmonik, K.; Patryn, J.; Iwanicka-Nowicka, R.; Biecek, P.; Wilczynski, B. *Arabidopsis* SWI/SNF chromatin remodeling complex binds both promoters and terminators to regulate gene expression. *Nucl. Acids Res.* **2017**, *45*, 3116–3129. [CrossRef]

129. Han, S.-K.; Sang, Y.; Rodrigues, A.; Wu, M.-F.; Rodriguez, P.L.; Wagner, D. The SWI2/SNF2 chromatin remodeling ATPase BRAHMA represses abscisic acid responses in the absence of the stress stimulus in *Arabidopsis*. *Plant Cell* **2012**, *24*, 4892–4906. [CrossRef]

130. Farrona, S.; Hurtado, L.; Reyes, J.C. A nucleosome interaction module is required for normal function of *Arabidopsis thaliana* BRAHMA. *J. Mol. Biol.* **2007**, *373*, 240–250. [CrossRef]

131. Li, C.; Gu, L.; Gao, L.; Chen, C.; Wei, C.Q.; Qiu, Q.; Chien, C.W.; Wang, S.; Jiang, L.; Ai, L.F.; et al. Concerted genomic targeting of H3K27 demethylase REF6 and chromatin-remodeling ATPase BRM in *Arabidopsis*. *Nat. Genet.* **2016**, *48*, 687–693. [CrossRef] [PubMed]

132. Peirats-Llobet, M.; Han, S.-K.; Gonzalez-Guzman, M.; Jeong, C.W.; Rodriguez, L.; Belda-Palazon, B.; Wagner, D.; Rodriguez, P.L. A direct link between abscisic Acid sensing and the chromatin-remodeling ATPase BRAHMA via core ABA signaling pathway components. *Mol. Plant* **2016**, *9*, 136–147. [CrossRef] [PubMed]

133. Kagale, S.; Rozwadowski, K. EAR motif-mediated transcriptional repression in plants: An underlying mechanism for epigenetic regulation of gene expression. *Epigenetics* **2011**, *6*, 141–146. [CrossRef] [PubMed]

134. Long, J.A.; Ohno, C.; Smith, Z.R.; Meyerowitz, E.M. TOPLESS regulates apical embryonic fate in *Arabidopsis*. *Science* **2006**, *312*, 1520–1523. [CrossRef]

135. Krogan, N.T.; Hogan, K.; Long, J.A. APETALA2 negatively regulates multiple floral organ identity genes in *Arabidopsis* by recruiting the co-repressor TOPLESS and the histone deacetylase HDA19. *Development* **2012**, *139*, 4180–4190. [CrossRef]

136. Oh, E.; Zhu, J.-Y.; Ryu, H.; Hwang, I.; Wang, Z.-Y. TOPLESS mediates brassinosteroid-induced transcriptional repression through interaction with BZR1. *Nat. Commun.* **2014**, *5*, 4140. [CrossRef]

137. Walter, J.; Nagy, L.; Hein, R.; Rascher, U.; Beierkuhnlein, C.; Willner, E.; Jentsch, A. Do plants remember drought? Hints towards a drought-memory in grasses. *Environ. Exp. Bot.* **2011**, *71*, 34–40. [CrossRef]

138. Ding, Y.; Fromm, M.; Avramova, Z. Multiple exposures to drought 'train' transcriptional responses in *Arabidopsis*. *Nat. Commun.* **2012**, *3*, 740. [CrossRef]

139. Virlouvet, L.; Fromm, M. Physiological and transcriptional memory in guard cells during repetitive dehydration stress. *New Phytol.* **2015**, *205*, 596–607. [CrossRef]

140. Ding, Y.; Liu, N.; Virlouvet, L.; Riethoven, J.-J.; Fromm, M.; Avramova, Z. Four distinct types of dehydration stress memory genes in *Arabidopsis thaliana*. *BMC Plant Biol.* **2013**, *13*, 229. [CrossRef]

141. Ding, Y.; Virlouvet, L.; Liu, N.; Riethoven, J.-J.; Fromm, M.; Avramova, Z. Dehydration stress memory genes of *Zea mays*; comparison with *Arabidopsis thaliana*. *BMC Plant Biol.* **2014**, *14*, 141. [CrossRef]

142. Virlouvet, L.; Avenson, T.J.; Du, Q.; Zhang, C.; Liu, N.; Fromm, M.; Avramova, Z.; Russo, S.E. Dehydration stress memory: Gene networks linked to physiological responses during repeated stresses of *Zea mays*. *Front. Plant Sci.* **2018**, *9*, 1058. [CrossRef] [PubMed]

143. Li, P.; Yang, H.; Wang, L.; Liu, H.; Huo, H.; Zhang, C.; Liu, A.; Zhu, A.; Hu, J.; Lin, Y.; et al. Physiological and transcriptome analyses reveal short-term responses and formation of memory under drought stress in rice. *Front. Genet.* **2019**, *10*, 55. [CrossRef]

144. Chen, Y.; Li, C.; Yi, J.; Yang, Y.; Lei, C.; Gong, M. Transcriptome response to drought, rehydration and re-dehydration in potato. *Int. J. Mol. Sci.* **2020**, *21*, 159. [CrossRef] [PubMed]

145. Kim, Y.-K.; Chae, S.; Oh, N.-I.; Nguyen, N.H.; Cheong, J.-J. Recurrent drought conditions enhance the induction of drought stress memory genes in *Glycine max* L. *Front. Genet.* **2020**, *11*, 576086. [CrossRef] [PubMed]

146. Ramírez, D.A.; Rolando, J.L.; Yactayo, W.; Monneveux, P.; Mares, V.; Quiroz, R. Improving potato drought tolerance through the induction of long-term water stress memory. *Plant Sci.* **2015**, *238*, 26–32. [CrossRef] [PubMed]

147. Wang, X.; Vignjevic, M.; Jiang, D.; Jacobsen, S.; Wollenweber, B. Improved tolerance to drought stress after anthesis due to priming before anthesis in wheat (*Triticum aestivum* L.) var. Vinjett. *J. Exp. Bot.* **2014**, *65*, 6441–6456. [CrossRef]

148. Tabassum, T.; Farooq, M.; Ahmad, R.; Zohaib, A.; Wahid, A.; Shahid, M. Terminal drought and seed priming improves drought tolerance in wheat. *Physiol. Mol. Biol. Plants* **2018**, *24*, 845–856. [CrossRef]

149. Abdallah, M.B.; Methenni, K.; Nouairi, I.; Zarrouk, M.; Youssef, N.B. Drought priming improves subsequent more severe drought in a drought-sensitive cultivar of olive cv. *Chétoui*. *Sci. Hortic.* **2017**, *221*, 43–52. [CrossRef]

150. Avramova, Z. Transcriptional 'memory' of a stress: Transient chromatin and memory (epigenetic) marks at stress-response genes. *Plant J.* **2015**, *83*, 149–159. [CrossRef]

151. Lämke, J.; Bäurle, I. Epigenetic and chromatin-based mechanisms in environmental stress adaptation and stress memory in plants. *Genome Biol.* **2017**, *18*, 124. [CrossRef] [PubMed]

152. Liu, N.; Ding, Y.; Fromm, M.; Avramova, Z. Different gene-specific mechanisms determine the 'revised-response' memory transcription patterns of a subset of *A. thaliana* dehydration stress responding genes. *Nucleic Acids Res.* **2014**, *42*, 5556–5566. [CrossRef]

153. Sani, E.; Herzyk, P.; Perrella, G.; Colot, V.; Amtmann, A. Hyperosmotic priming of *Arabidopsis* seedlings establishes a long-term somatic memory accompanied by specific changes of the epigenome. *Genome Biol.* **2013**, *14*, R59. [CrossRef] [PubMed]

154. Boyko, A.; Kovalchuk, I. Genome instability and epigenetic modification–heritable responses to environmental stress? *Curr. Opin. Plant Biol.* **2011**, *14*, 260–266. [CrossRef] [PubMed]

155. Springer, N.M. Epigenetics and crop improvement. *Trends Genet.* **2013**, *29*, 241–247. [CrossRef] [PubMed]

156. Mickelbart, M.V.; Hasegawa, P.M.; Bailey-Serres, J. Genetic mechanisms of abiotic stress tolerance that translate to crop yield stability. *Nat. Rev. Genet.* **2015**, *16*, 237–251. [CrossRef] [PubMed]

157. Bilichak, A.; Kovalchuk, I. Transgenerational response to stress in plants and its application for breeding. *J. Exp. Bot.* **2016**, *67*, 2081–2092. [CrossRef] [PubMed]

158. Raju, S.K.K.; Shao, M.-R.; Sancheza, R.; Xu, Y.-Z.; Sandhub, A.; Graef, G.; Mackenziea, S. An epigenetic breeding system in soybean for increased yield and stability. *Plant Biotechnol. J.* **2018**, *16*, 1836–1847. [CrossRef]

159. Verkest, A.; Byzova, M.; Martens, C.; Willems, P.; Verwulgen, T.; Slabbinck, B.; Rombaut, D.; Van de Velde, J.; Vandepoele, K.; Standaert, E.; et al. Selection for improved energy use efficiency and drought tolerance in canola results in distinct transcriptome and epigenome changes. *Plant Physiol.* **2015**, *168*, 1338–1350. [CrossRef]

160. Tabassum, T.; Farooq, M.; Ahmad, R.; Zohaib, A.; Wahid, A. Seed priming and transgenerational drought memory improves tolerance against salt stress in bread wheat. *Plant Physiol. Biochem.* **2017**, *118*, 362–369. [CrossRef]

161. Zheng, X.; Chen, L.; Xia, H.; Wei, H.; Lou, Q.; Li, M.; Li, T.; Luo, L. Transgenerational epimutations induced by multi-generation drought imposition mediate rice plant's adaptation to drought condition. *Sci. Rep.* **2017**, *7*, 39843. [CrossRef] [PubMed]

162. Crisp, P.; Ganguly, D.; Eichten, S.R.; Borevitz, J.O.; Pogson, B.J. Reconsidering plant memory: Intersections between stress recovery, RNA turnover, and epigenetics. *Sci. Adv.* **2016**, *2*, e1501340. [CrossRef] [PubMed]

Publisher's Note: MDPI stays neutral with regard to jurisdictional claims in published maps and institutional affiliations.

International Journal of
Molecular Sciences

Review

Mediator Complex: A Pivotal Regulator of ABA Signaling Pathway and Abiotic Stress Response in Plants

Leelyn Chong, Pengcheng Guo and Yingfang Zhu * 

State Key Laboratory of Crop Stress Adaptation and Improvement, School of Life Sciences, Henan University, Kaifeng 475001, China; chong87@henu.edu.cn (L.C.); gpc0108@henu.edu.cn (P.G.)
* Correspondence: zhuyf@henu.edu.cn

Received: 25 September 2020; Accepted: 19 October 2020; Published: 20 October 2020

Abstract: As an evolutionarily conserved multi-protein complex, the Mediator complex modulates the association between transcription factors and RNA polymerase II to precisely regulate gene transcription. Although numerous studies have shown the diverse functions of Mediator complex in plant development, flowering, hormone signaling, and biotic stress response, its roles in the Abscisic acid (ABA) signaling pathway and abiotic stress response remain largely unclear. It has been recognized that the phytohormone, ABA, plays a predominant role in regulating plant adaption to various abiotic stresses as ABA can trigger extensive changes in the transcriptome to help the plants respond to environmental stimuli. Over the past decade, the Mediator complex has been revealed to play key roles in not only regulating the ABA signaling transduction but also in the abiotic stress responses. In this review, we will summarize current knowledge of the Mediator complex in regulating the plants' response to ABA as well as to the abiotic stresses of cold, drought and high salinity. We will particularly emphasize the involvement of multi-functional subunits of MED25, MED18, MED16, and CDK8 in response to ABA and environmental perturbation. Additionally, we will discuss potential research directions available for further deciphering the role of Mediator complex in regulating ABA and other abiotic stress responses.

Keywords: Mediator complex; transcription; ABA signaling; abiotic stress response

1. Introduction

The Mediator is an evolutionarily conserved eukaryotic multi-protein complex that has been recognized as a key regulator of plant growth and development, plant defense, and hormone signaling transduction [1–3]. It regulates transcription through recruiting RNA polymerase II (Pol II) to specific gene promoters by linking transcription factors (TFs) bound at activators and repressors with the pre-initiation complex (PIC). Upon receiving and transferring regulatory signals to the basal transcriptional machinery, the Mediator complex undergoes conformational changes, which creates a flexible surface that aids the assembly of PIC. Functioning as a molecular bridge, the Mediator complex physically interacts with PIC as well as TFs to perform transcriptional activation [4–6]. Based on the classification from structural studies, the core Mediator is divided into the head, middle, and tail modules [7–9]. Each of these modules is made up of different subunits that characterize the distinct function of each module on transcription [10]. Depending on the species, the number of Mediator subunits may vary, and there are approximately 34 subunits reported in plant Mediator [11]. A number of Mediator subunits has already been revealed to have critical functions in various plant developmental processes, hormone signaling, plant defense, and abiotic stress tolerance [1,3,12,13]. The head module primarily associates with Pol II to affect transcription whereas the tail module is believed to play a highly significant role as it interacts with gene-specific TFs. The middle module is reported to be responsible for the transfer of transcription signal from the tail to the head, which may also interact with Pol II [14]. The fourth and separable kinase module, termed as the CDK8 module, which consists of CDK8, C-type cyclin (CycC), MED12, and MED13 subunits, has been indicated to exist in plants. *Arabidopsis* CDK8 was first reported to regulate floral organ identity [15]. It was later found to interact with MED14 and *Arabidopsis* LEUNIG, a transcription co-repressor [16]. Further studies on the *Arabidopsis* regulator of alternative oxidase 1 (*rao1*) mutant that carries a mutation in *CDK8* documented that *CDK8* regulates mitochondrial retrograde signaling under H_2O_2 and cold stress [17].

Due to the important role of Mediator complex in transcription, it is comprehensible to find a rise of studies revealing the engagement of Mediator complex in various responses to ABA and environmental disturbances such as biotic and abiotic stresses. The role of Mediator complex in response to biotic stresses has been well documented [1,3,18–20]. However, the function of Mediator in the context of ABA and abiotic stress response still require further investigation as only a few studies have examined the role of Mediator in responding to ABA as well as to environmental perturbation. Thus far, the Mediator complex has only been discovered to serve roles in cold, salt, and drought stresses. Despite these findings, more studies are still required to facilitate the research in this area. Therefore, this review will emphasize the regulatory roles of Mediator complex in the ABA signaling pathway, as it is a major phytohormone that contributes significantly to the plant's ability in adapting abiotic stress. Furthermore, we will also discuss the most recently reported role of Mediator complex in three abiotic stresses of cold, high salinity, and drought.

2. The Importance of Mediator Complex in Transcriptional Regulation

The Mediator complex functions together with cofactors of Pol II to regulate gene expression at the transcriptional level. Mediator plays a significant role in assisting plants to adapt to environmental changes because TFs recruit the Mediator complex through protein–protein interaction to trigger the activation or repression of target genes in plants with the Pol II transcription complex [1]. A diverse range of biological processes in *Arabidopsis* is regulated by more than 1600 TFs [21]. TFs are linked with the Mediator complex since they are crucial components in the Pol II-based transcriptional machinery and the Mediator complex interacts with different TFs upon conformational changes or when environmental and cellular signals are perceived. More specifically, the subunits of the Mediator complex are the vital components that interact with the TFs to regulate transcription. In fact, 34 subunits of the Mediator complex that have been purified from *Arabidopsis* were reported thus far to have the possibility of interacting with different TFs [2]. In terms of ABA signaling and abiotic stresses, the Mediator's role and regulation of gene(s) induced by each abiotic stress situation and ABA require further work in order to be fully understood. Hence, it is important to identify any potential interaction that may occur between Mediator subunits and TFs that are involved in the signaling pathway of each situation of the abiotic stress as well as in ABA signaling.

3. Mediator Complex as a Pivotal Regulator of ABA Signaling Pathway

ABA is a phytohormone that has profound functions in various developmental processes throughout the plant life cycle, such as seed germination and dormancy, organ size control, vegetative development, stomatal closure regulation, as well as senescence [22–25]. It has been reported that the concentrations of ABA can increase up to 50-fold under drought stress [26] and this is one of the most drastic changes observed thus far in the concentration of a plant hormone responding to an environmental stimulus. Due to ABA's significant involvement in plants' responses to various environmental stresses, the ABA signaling pathway has been studied extensively. Thus far, many of the key components of the pathway have been successfully identified [27]. Despite this, components in the downstream of ABA signaling pathway remain to be uncovered.

Since 2009, a group of PYRABACTIN RESISTANCE (PYR)/PYR1-LIKE (PYL)/Regulatory Components of ABA Receptor (RCAR) proteins, members of a family of 14 START-domain-containing proteins in *Arabidopsis*, have been shown to function as the ABA receptors [28,29]. The core ABA signaling pathway also consists of the protein kinases in the SNF1-related protein kinase 2 (SnRK2) family, particularly SnRK2.2, SnRK2.3, and SnRK2.6/Open Stomata 1 (OST1) [27]. They have been shown to function as key positive regulators of ABA signaling [30–33]. Therefore, the earliest events occurred in ABA signaling require the presence of PYR/PYL/RCAR proteins, PP2Cs, and SnRK2 kinases (as shown in Figure 1). Without ABA, PP2Cs represses the kinase activity of SnRK2s as well as the downstream ABA signaling events. In the presence of ABA, it induces the formation of PYRs/PYLs/RCARs-ABA-PP2Cs complexes and PP2Cs will become inactivated, thereby permitting SnRK2s activation and the downstream events of ABA signaling [23,25,27,34]. The core ABA signaling pathway has been reconstituted successfully with those key components in vitro [35]. Interestingly, several recent studies simultaneously showed that Raf-like kinases (RAFs) could quickly activate SnRK2s to respond to ABA, osmotic and drought stress by direct phosphorylation [36–39].

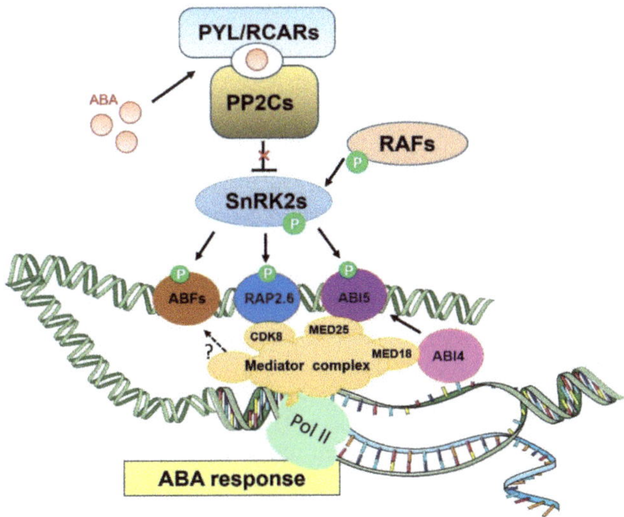

Figure 1. The pivotal role of Mediator complex in the ABA signaling pathway. ABA is perceived by its receptors PYL/RCARs, which promotes the interaction between PP2Cs (negative regulators of the ABA signaling pathway) and PYLs, hence releasing the positive regulators SnRK2s to activate ABA downstream signaling events. Additionally, RAFs can directly phosphorylate SnRK2s for the activation of SnRK2s, which subsequently interact with and phosphorylate several downstream TFs including ABFs, ABI5 and RAP2.6 to transduce the ABA signals. Mediator subunits of CDK8, MED25, and MED18 relay the signals from TFs RAP2.6, ABI5, and ABI4, respectively, and help recruit the RNA Pol II to the TFs-targeted promoters of ABA-responsive genes, thereby promoting the transcription of ABA-responsive genes.

The Mediator complex as described is a critical co-regulator of the transcriptional machinery and, unsurprisingly, it has also been found to serve important roles in the ABA signaling transduction. In fact, *MED25* is the first Mediator subunit that has been reported to act in response to ABA [40]. It was found that *MED25* negatively regulates the ABA signaling pathway as *med25* mutants display an increased sensitivity to ABA during seed germination and early seedling growth [40]. Consistent with its negative role in ABA signaling, *med25* mutant was noted to have an increased expression of ABA-responsive genes in response to ABA treatment compared to the wild type (WT) plants. ABA induced the transcription of *ABI5* (*ABA-INSENSITIVE5*), a key TF regulating the ABA signaling during seed germination [41–43], and, intriguingly, the ABA-induced transcription of *ABI5* was suppressed in *med25* mutants compared to WT. Nevertheless, ABI5 protein accumulated at higher abundance in *med25* mutants than that in WT, implying that MED25 may negatively regulate ABI5 at post-transcriptional level. Chromatin immunoprecipitation (ChIP) experiments further indicated that MED25 was highly enriched at the promoters of *ABI5* downstream genes, and this enrichment was reduced upon ABA treatment. MED25 was shown to directly interact with ABI5 and this interaction was attenuated by ABA, which was in accordance with the negative impacts of MED25 on the ABI5-regulated ABA responses. It is worth noting that *MED25* may be a critical regulator in hormones crosstalk between Jasmonic acid (JA), ethylene, and ABA signaling due to its interaction with MYC2 and several TFs in plants [44–46]. The head module subunit *MED18* has also been implicated in the ABA signaling. Opposite to *med25*, *med18* mutants are more insensitive to ABA at seed germination and early growth stages, similar to *abi4* and *abi5* mutants [47]. Remarkably, the induced expression of *ABI4* and *ABI5* by ABA are much lower in *med18* mutants than those in WT, indicating that the transcription of *ABI4* and *ABI5* are positively regulated by MED18. ChIP-qPCR revealed that MED18 is recruited to the ABI4 binding site on the *ABI5* promoter under both mock and ABA treatments. The physical interaction between MED18 and TF ABI4 further supports that MED18 regulates the ABA response and expression of *ABI5* through interacting with ABI4.

Recently, another subunit belonging to the Mediator kinase module termed as *CDK8*, has been identified as a critical regulator in the ABA signaling pathway [48]. As described previously, SnRK2s need to be phosphorylated by certain protein kinases in order to further perform the ABA signaling process [36]. CDK8 is known to possess kinase activity and this presents an opportunity for exploring its potential in regulating SnRK2s. Through utilizing genetic, transcriptomic, and biochemical approaches, CDK8 was solidified to associate with RAP2.6 and SnRK2.6 to positively regulate the transcription of ABA-responsive genes. *CDK8* mutation led to ABA insensitivity. Conversely, *CDK8* over-expression lines displayed hypersensitivity to ABA. Interestingly, the kinase-inactive version of CDK8 did not rescue the ABA phenotype of *cdk8* mutants, indicating the requirement of CDK8 kinase activity in the ABA response. The CDK8 and its kinase module components are generally known as negative regulators of gene expression in yeast, metazoan cells, and plants [13,49,50]. However, increasing evidence is showing that CDK8 could also play a positive role in plant transcriptional regulation as expression of defense-responsive genes (*PDF1.2*, *AACT1* and *NPR1*), salicylic acid (SA)-biosynthetic genes (*ICS1* and *EDS5*) and ABA-responsive genes such as *RAP2.6*, *RD29A*, *RD29B*, and *COR15A* [44,48,51,52] are positively regulated by *CDK8* in plants. Transcriptomic analysis has revealed that *CDK8* affects approximately 30% of the ABA-responsive genes, most of these genes are downregulated in *cdk8* mutants compared to WT. The expression of several important TFs (*DREB2A* and *RAP2.6*) and ABA-responsive genes (*RD29A*, *RD29B*, and *COR15A*) was found to be significantly lower in *cdk8* mutant plants. Therefore, this indicates a positive role of *CDK8* in modulating ABA-induced transcription. Moreover, ChIP analysis was utilized to verify that *CDK8* is essential for the ABA-induced Pol II recruitment to the promoters of ABA-responsive genes. In fact, RAP2.6, an ERF/AP2 type TF that involves in biotic and abiotic stress responses, was identified as a new interactor of CDK8 through a yeast two-hybrid screen. CDK8 was further shown to be enriched at the promoter region of *RAP2.6* in response to ABA, demonstrating that *CDK8* is an important component for regulating *RAP2.6* transcription. Moreover, *RAP2.6* was found to directly associate with the DRE or GCC motif and *RD29A* or *COR15A* promoters. In response to ABA, RAP2.6 could be enriched at the *RD29A* and *COR15A* promoters. These findings indicated the possibility that *CDK8* may regulate the expression of ABA-responsive genes through *RAP2.6* [48]. It may also be possible that other TFs interact with CDK8 to regulate the expression of ABA-responsive genes.

In addition, RAP2.6-mediated activation of *RD29A* has been observed to be attenuated in *cdk8* mutants, thereby showing that *CDK8* is required for the recruitment of Pol II to the promoters of RAP2.6 target genes. Consistent with biochemical results, the over-expression of *RAP2.6* resulted in hypersensitivity to ABA and mannitol as well as higher expressions of several ABA-responsive genes. These findings indicated that RAP2.6 and CDK8 could finetune the transcription of ABA-responsive genes, especially those genes containing DRE/GCC-motifs. Another important finding is that Mediator

CDK8 could link the core ABA signaling component of SnRK2.6 to Pol II transcriptional machinery, which facilitates the immediate transcriptional response to ABA and abiotic stress. Although no direct interaction and phosphorylation between CDK8 and SnRK2.6 have been observed, it is possible that CDK8 associates with SnRK2.6 through RAP2.6 to form a ternary complex since both kinases directly interact with RAP2.6. In vitro kinase assays further indicated that RAP2.6 was phosphorylated by SnRK2.6, but not by CDK8. It therefore raises the possibility that RAP2.6 may act as a SnRK2.6 substrate or a downstream TF to transduce the ABA signaling, but it requires further genetic studies and in vivo phosphorylation evidence to support the existence of this ternary complex in plants. Future study should also elucidate whether the phosphorylation of RAP2.6 by SnRK2.6 could affect its transcriptional activity, protein stability or translocation. Although CDK8 did not directly phosphorylate RAP2.6 in vitro, the possibility of CDK8 kinase activity requiring either cyclin or other partners to promote its phosphorylation in vivo should not be excluded. Thus far, very few CDK8 substrates have been reported and this is likely due to its weak kinase activity in vitro. The pivotal roles of Mediator complex in the ABA signaling pathway are summarized in Figure 1.

4. Mediator Complex Is Vital for Plants to Respond to Abiotic Stresses

In order to withstand disturbances in the natural environment, plants must be able to rapidly respond and adapt to environmental stimuli by dynamically changing the expression of genes that help them maintain cellular homeostasis. Abiotic stresses such as cold, high salinity, and drought are some of the environmental stimuli that plants get exposed to and they must be able to integrate these signals using different regulatory pathways if they are to survive [25,53]. Various subunits of the Mediator complex including CDK8, MED16, MED14, and MED25 have been identified to help plants to respond to these stresses [1,48]. We will summarize some of the findings that have been reported about the functions of these subunits in dealing with three abiotic stresses of cold, high salinity, and drought (Figure 2).

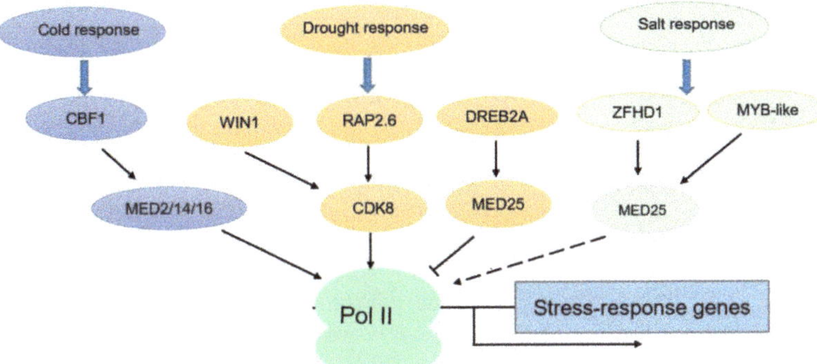

Figure 2. The simplified network of TFs and Mediator complex in regulating the abiotic stress responses. In response to cold stress, CBF1 activates the expression of *COR* genes through MED2, MED14, and MED16, which are required for the recruitment of RNA Pol II to the promoters of *COR* genes; In response to drought stress, CDK8 physically interact with WIN1 and RAP2.6 to positively regulate the cuticle wax biosynthesis and expression of stress-responsive genes; in contrast, MED25 negatively regulates the transcriptional activity of DREB2A and the expression of stress-responsive genes, thereby negatively contributing to the drought tolerance; in response to salt stress, ZFHD1 and MYB-like interact with MED25 to positively regulate the salt response.

5. Mediator Subunits Modulate Freezing Tolerance in Plants

MED16 is one of the first Mediator subunits that was reported to involve in abiotic stress response. *MED16* has been indicated to help plants overcome cold stress (freezing) through eliciting responses that maintain physiological metabolic homeostasis. Before MED16 was recognized as part of the Mediator complex, it was named as *SENSITIVE TO FREEZING6* (*SFR6*) and was identified for its role in cold acclimation-induced freezing tolerance [54,55]. The process of cold acclimatization involves the expression of many cold inducible/cold responsive or cold on-regulated (*COR*) genes such as *KIN1, COR15a*, and *RD29A* (specifically those consisting of C-repeat/dehydration-responsive element (CRT/DRE) elements in their promoter). The expression of *COR* genes is mainly induced by the TF of C-Repeat/DRE Binding Factor 1 (CBF1) [56]. In the study for its ability of freezing tolerance and cold acclimation, the *sfr6* mutants were observed to express significantly decreased levels of *COR* gene and protein accumulation and were thereby unable to tolerate freezing after cold acclimation. The CRT/DRE elements containing *COR* genes become uninducible at low temperature in *sfr6* mutants. It is very likely that SFR6/MED16 acts downstream of CBF1 and triggers the recruitment of Mediator complex to CBF1 responsive genes. To better understand the role of *MED16* in cold signaling, TFs *CBF1* and *CBF2* were also overexpressed in the *sfr6* mutant since CBFs are responsible for the activation of *COR* genes. It was found that the overexpression of *CBF1* and *CBF2* failed to increase the expression of *COR* target genes, further confirming that *MED16* acts downstream of CBF TFs [57]. In fact, *MED16* has been validated as an indispensable Mediator subunit in plants for activating CBF-regulated *COR* genes as, without MED16, Pol II is unable to be recruited to these genes [58]. The plant's ability to survive cold stress relies on the effective induction of *COR* genes and interestingly, without *MED16*, *COR* genes are unable to be induced, and this further causes osmotic stress sensitivity in *sfr6* mutants [55]. The *med16* mutant was also reported for its hypersensitivity to iron deficiency and sensitivity to excessive zinc, which could be rescued by increasing iron concentration. Additionally, MED16 was proven to interact with MED25 to regulate iron homeostasis [59]. Despite its association with MED16 to regulate biological processes, *MED25* was not involved in cold acclimation-induced freezing tolerance.

In addition to *MED16*, *MED14* and *MED2* are two additional Mediator subunits that have been shown to have an effect on *COR* gene expression, further signifying the importance of Mediator in plant's adaptation to cold stress [58]. More importantly, all three tail module subunits of MED16, MED14, and MED2 play a significant role in recruiting the Pol II to the CBF1 target genes to regulate cold stress response [58], suggesting the essential role of Mediator complex in the cold response.

6. Multi-Functional Roles of Mediator in Salt and Drought Stresses

Salt and drought are two major abiotic stresses that limit crop yield worldwide. The SOS (Salt Overly Sensitive) signaling pathway is extensively reported to contribute to salt tolerance in plants [53,60]. The transcription of TFs is also essential for the salt and drought response in plants [61]. Thus far, only *med25* and *med18* mutants have been reported to exhibit a reduced tolerance to salt stress [45,62]. MED25 was found to interact with several TFs of DREB2A (drought response element protein B), ZFHD1 (zinc finger homeodomain 1) and MYB-like from yeast two-hybrid screen using the conserved activator-interacting domain (ACID) of MED25 as a bait. Consistently, mutation in *MED25*, *DREB2A, ZFHD1*, and *MYB-like* all caused an increased sensitivity to salt stress [45]. Nevertheless, the salt-responsive genes that are affected by *MED25* and those TFs are not reported. MED18 was found to interact with NUP85 and positively contribute to the ABA signaling and salt tolerance [62]. A recent work also reported that four Mediator subunits (*MED9, MED16, MED18*, and *CDK8*), representing four different modules, are required for salt stress and thermal stress mediated transcriptional responses by RNA-seq analysis in *Arabidopsis* [63]. However, limited studies have been reported on the roles of Mediator complex in salt stress responses. The detailed mechanism of how Mediator subunits regulate the salt response remains unclear. It is unknown if Mediator complex could affect the SOS pathways and any other critical transporters.

Besides salt stress, *MED25* is also involved in drought stress. The *med25* mutant has been indicated to display an increased resistance to drought, as opposed to its salt sensitivity [45]. MED25 is involved in modulating drought stress response through interacting with DREB2A. DREB2A consists of both repressing domain (RD) and activating domain (AD) in its protein sequence [64,65]. The mutation of *dreb2a* and *med25* has been demonstrated to have an opposite effect in drought stress as *dreb2a* was found to exhibit drought sensitivity while *med25* displayed an increased resistance to drought. The explanation provided for the observed opposite effect of *MED25* and *DREB2A* in drought stress was that MED25 acts as the corepressor of DREB2A in drought stress by interacting with the AD in DREB2A and depositing some other Mediator subunit in close vicinity of DREB2A RD. Therefore, when *MED25* is disrupted, the repressor function is lost and DREB2A activates genes involved in drought [45]. Based on the evidence presented about *MED25*, it appears that *MED25* mainly plays negative roles in abiotic stress response.

In addition, *CDK8* has also been indicated to participate in drought stress recently. *CDK8* mutation results in higher stomatal density and impaired stomatal aperture, as well as reduced tolerance to drought [48]. Consistently, over-expression of *CDK8* enhances the drought tolerance. Considering the enhanced cuticle permeability and thinner cutin observed in *cdk8* mutants [44], it is likely that *CDK8* regulates the drought response through multiple mechanisms. Remarkably, CDK8 was found to directly interact with ERF/AP2 type TFs WIN1 (WAX INDUCER1) and RAP2.6, which are key regulators of cuticle wax biosynthesis and an abiotic stress responsive gene, respectively [44,66]. CDK8 positively regulates cutin biosynthesis and wax accumulation through interacting with WIN1. Interestingly, in addition to playing a role in the wax biosynthesis, WIN1 also participates in abiotic stress response as its expression is significantly induced by various abiotic stresses and WIN1 can also bind the GCC-box and DRE element sequences to activate several stress-responsive genes [67,68], implying the potential function of CDK8–WIN1 interaction in the drought response. Furthermore, CDK8 also contributes to drought tolerance by cooperating with RAP2.6-SnRK2.6 complex, which could facilitate the immediate transcription of stress-responsive genes. Therefore, Mediator subunits are capable of different functions and can perform different roles depending on the type of environmental stress.

7. Conclusion and Perspectives

In response to abiotic stress, plants must appropriately regulate gene expression in a synchronized manner. It is unsurprising to find that the Mediator complex is linked with the ABA signaling pathway and abiotic stress as it has important roles in transcriptional regulation. Despite the confirmed relationship of Mediator complex with ABA and abiotic stress response, the molecular mechanism of Mediator complex in regulating the ABA and abiotic stress response remains elusive. Thus far, only a few Mediator subunits have been reported to be involved in the ABA signaling pathway, cold (freezing), salt, and drought response. Since more than 30 Mediator subunits have been documented in plants, this presents an opportunity for discovering if there are more subunits involving in the ABA signaling and abiotic stress response in future. Furthermore, the plausible roles of Mediator complex in heat stress and submergence are worthy of an investigation as knowledge about the mechanism of heat and submergence stress response is still limited. Therefore, it is desirable to screen all the Mediator subunits and identify the ones that exhibit functions in the ABA signaling pathway as well as in abiotic stress that has not yet been fully studied.

As revealed from structural studies, the Mediator complex is divided into four distinct modules and it is still unclear whether each module could exert specific effects on the ABA signaling transduction or abiotic stress response in plants. Future studies should address whether the subunits within the same module present overlapping or opposite roles in regulating the ABA and abiotic stress responses. It is known that *med25* mutants are sensitive to ABA, while the *cdk8* and *med18* mutants are even less sensitive to ABA. It is necessary to study the detailed mechanism of how those Mediator subunits coordinately or completely regulate the ABA or abiotic stress. Given the nature that Mediator complex functions between Pol II and TFs, it is also necessary to identify additional TFs that interact with different Mediator subunits in response to ABA and abiotic stress. Currently, only a few TFs (DREB2A, ABI5 and RAP2.6, etc.) have been reported to interact with Mediator subunits to regulate ABA and abiotic stress responses. Undoubtedly, this area is drawing the attention of plant scientists as the Mediator complex profoundly participates in transcriptional regulation. High-throughput proteomics and protein–protein interaction approaches are essential for improving the knowledge of this field and they should be further improvised to uncover more TFs that directly interact with specific Mediator subunits, which will then provide deep insights into the molecular mechanism of Mediator complex in the regulation of ABA signaling and abiotic stress. Furthermore, it would be interesting to find if there are potential ABA or stress-induced dynamic interactions between MED and TFs in response to a specific environmental stimulus.

Author Contributions: L.C., Y.Z. and P.G. wrote the manuscript; Y.Z. edited the manuscript. All authors have read and agreed to the published version of the manuscript.

Funding: This work was funded by NSFC 31900238 to Y.Z.

Acknowledgments: We apologize to colleagues whose work could not be cited due to space limitations.

Conflicts of Interest: The authors declare no conflict of interest.

References

1. Yang, Y.; Li, L.; Qu, L.-J. Plant Mediator complex and its critical functions in transcription regulation. *J. Integr. Plant Biol.* **2015**, *58*, 106–118. [CrossRef] [PubMed]
2. Kidd, B.N.; Cahill, D.M.; Manners, J.M.; Schenk, P.M.; Kazan, K. Diverse roles of the Mediator complex in plants. *Semin. Cell Dev. Biol.* **2011**, *22*, 741–748. [CrossRef] [PubMed]
3. Zhai, Q.; Li, C. The plant Mediator complex and its role in jasmonate signaling. *J. Exp. Bot.* **2019**, *70*, 3415–3424. [CrossRef] [PubMed]
4. Schilbach, S.; Hantsche, M.; Tegunov, D.; Dienemann, C.; Wigge, C.; Urlaub, H.; Cramer, P. Structures of transcription pre-initiation complex with TFIIH and Mediator. *Nat. Cell Biol.* **2017**, *551*, 204–209. [CrossRef]
5. Roeder, R.G. The role of general initiation factors in transcription by RNA polymerase II. *Trends Biochem. Sci.* **1996**, *21*, 327–335. [CrossRef]
6. Kornberg, R.D. Mediator and the mechanism of transcriptional activation. *Trends Biochem. Sci.* **2005**, *30*, 235–239. [CrossRef]
7. Plaschka, C.; Larivière, L.; Wenzeck, L.; Seizl, M.; Hemann, M.; Tegunov, D.; Petrotchenko, E.V.; Borchers, C.H.; Baumeister, W.; Herzog, F.; et al. Architecture of the RNA polymerase II–Mediator core initiation complex. *Nat. Cell Biol.* **2015**, *518*, 376–380. [CrossRef]
8. Dotson, M.R.; Yuan, C.X.; Roeder, R.G.; Myers, L.C.; Gustafsson, C.M.; Jiang, Y.W.; Li, Y.; Kornberg, R.D.; Asturias, F.J. Structural organization of yeast and mammalian mediator complexes. *Proc. Natl. Acad. Sci. USA* **2000**, *97*, 14307–14310. [CrossRef]
9. Taatjes, D.J.; Näär, A.M.; Iii, F.A.; Nogales, E.; Tjian, R.; Andel, F. Structure, Function, and Activator-Induced Conformations of the CRSP Coactivator. *Science* **2002**, *295*, 1058–1062. [CrossRef]
10. Robinson, P.J.; Trnka, M.J.; Bushnell, D.A.; Davis, R.E.; Mattei, P.-J.; Burlingame, A.L.; Kornberg, R.D. Structure of a Complete Mediator-RNA Polymerase II Pre-Initiation Complex. *Cell* **2016**, *166*, 1411–1422.e16. [CrossRef]
11. Mathur, S.; Vyas, S.; Kapoor, S.; Tyagi, A.K. The Mediator Complex in Plants: Structure, Phylogeny, and Expression Profiling of Representative Genes in a Dicot (Arabidopsis) and a Monocot (Rice) during Reproduction and Abiotic Stress. *Plant Physiol.* **2011**, *157*, 1609–1627. [CrossRef] [PubMed]
12. Bäckström, S.; Elfving, N.; Nilsson, R.; Wingsle, G.; Björklund, S. Purification of a Plant Mediator from Arabidopsis thaliana Identifies PFT1 as the Med25 Subunit. *Mol. Cell* **2007**, *26*, 717–729. [CrossRef] [PubMed]
13. Malik, N.; Agarwal, P.; Tyagi, A. Emerging functions of multi-protein complex Mediator with special emphasis on plants. *Crit. Rev. Biochem. Mol. Biol.* **2017**, *52*, 475–502. [CrossRef]
14. Chadick, J.Z.; Asturias, F.J. Structure of eukaryotic Mediator complexes. *Trends Biochem. Sci.* **2005**, *30*, 264–271. [CrossRef] [PubMed]
15. Wang, W.-M.; Chen, X. HUA ENHANCER3 reveals a role for a cyclin-dependent protein kinase in the specification of floral organ identity in Arabidopsis. *Development* **2004**, *131*, 3147–3156. [CrossRef] [PubMed]
16. Gonzalez, D.; Bowen, A.J.; Carroll, T.S.; Conlan, R.S. The Transcription Corepressor LEUNIG Interacts with the Histone Deacetylase HDA19 and Mediator Components MED14 (SWP) and CDK8 (HEN3) To Repress Transcription. *Mol. Cell. Biol.* **2007**, *27*, 5306–5315. [CrossRef] [PubMed]

17. Ng, S.; Giraud, E.; Duncan, O.; Law, S.R.; Wang, Y.; Xu, L.; Narsai, R.; Carrie, C.; Walker, H.; Day, D.A.; et al. Cyclin-dependent Kinase E1 (CDKE1) Provides a Cellular Switch in Plants between Growth and Stress Responses. *J. Biol. Chem.* **2012**, *288*, 3449–3459. [CrossRef] [PubMed]

18. Dhawan, R.; Luo, H.; Foerster, A.M.; AbuQamar, S.; Du, H.-N.; Briggs, S.D.; Scheid, O.M.; Mengiste, T. HISTONE MONOUBIQUITINATION1 Interacts with a Subunit of the Mediator Complex and Regulates Defense against Necrotrophic Fungal Pathogens in Arabidopsis. *Plant Cell* **2009**, *21*, 1000–1019. [CrossRef]

19. Zhang, X.; Wang, C.; Zhang, Y.; Sun, Y.; Mou, Z. The Arabidopsis Mediator Complex Subunit16 Positively Regulates Salicylate-Mediated Systemic Acquired Resistance and Jasmonate/Ethylene-Induced Defense Pathways. *Plant Cell* **2012**, *24*, 4294–4309. [CrossRef]

20. An, C.; Mou, Z. The function of the Mediator complex in plant immunity. *Plant Signal. Behav.* **2013**, *8*, e23182. [CrossRef]

21. Qu, L.-J.; Zhu, Y.-X. Transcription factor families in Arabidopsis: Major progress and outstanding issues for future research. *Curr. Opin. Plant Biol.* **2006**, *9*, 544–549. [CrossRef]

22. Weiner, J.J.; Peterson, F.C.; Volkman, B.F.; Cutler, S.R. Structural and functional insights into core ABA signaling. *Curr. Opin. Plant Biol.* **2010**, *13*, 495–502. [CrossRef]

23. Cutler, S.R.; Rodriguez, P.L.; Finkelstein, R.R.; Abrams, S.R. Abscisic Acid: Emergence of a Core Signaling Network. *Annu. Rev. Plant Biol.* **2010**, *61*, 651–679. [CrossRef]

24. Finkelstein, R.R.; Gibson, S.I. ABA and sugar interactions regulating development: Cross-talk or voices in a crowd? *Curr. Opin. Plant Biol.* **2002**, *5*, 26–32. [CrossRef]

25. Zhu, J.-K. Abiotic Stress Signaling and Responses in Plants. *Cell* **2016**, *167*, 313–324. [CrossRef] [PubMed]

26. Zeevaart, J.A.D. Changes in the Levels of Abscisic Acid and Its Metabolites in Excised Leaf Blades of Xanthium strumarium during and after Water Stress. *Plant Physiol.* **1980**, *66*, 672–678. [CrossRef]

27. Chen, K.; Li, G.; Bressan, R.A.; Song, C.; Zhu, J.; Zhao, Y. Abscisic acid dynamics, signaling, and functions in plants. *J. Integr. Plant Biol.* **2020**, *62*, 25–54. [CrossRef] [PubMed]

28. Ma, Y.; Szostkiewicz, I.; Korte, A.; Moes, D.; Yang, Y.; Christmann, A.; Grill, E. Regulators of PP2C Phosphatase Activity Function as Abscisic Acid Sensors. *Science* **2009**, *324*, 1064–1068. [CrossRef]

29. Park, S.-Y.; Fung, P.; Nishimura, N.; Jensen, D.R.; Fujii, H.; Zhao, Y.; Lumba, S.; Santiago, J.; Rodrigues, A.; Chow, T.-F.F.; et al. Abscisic Acid Inhibits Type 2C Protein Phosphatases via the PYR/PYL Family of START Proteins. *Science* **2009**, *324*, 1068–1071. [CrossRef] [PubMed]

30. Fujii, H.; Verslues, P.E.; Zhu, J.-K. Identification of Two Protein Kinases Required for Abscisic Acid Regulation of Seed Germination, Root Growth, and Gene Expression in Arabidopsis. *Plant Cell* **2007**, *19*, 485–494. [CrossRef]

31. Fujii, H.; Zhu, J.-K. Arabidopsis mutant deficient in 3 abscisic acid-activated protein kinases reveals critical roles in growth, reproduction, and stress. *Proc. Natl. Acad. Sci. USA* **2009**, *106*, 8380–8385. [CrossRef] [PubMed]

32. Mustilli, A.-C.; Merlot, S.; Vavasseur, A.; Fenzi, F.; Giraudat, J. Arabidopsis OST1 Protein Kinase Mediates the Regulation of Stomatal Aperture by Abscisic Acid and Acts Upstream of Reactive Oxygen Species Production. *Plant Cell* **2002**, *14*, 3089–3099. [CrossRef]

33. Nakashima, K.; Fujita, Y.; Kanamori, N.; Katagiri, T.; Umezawa, T.; Kidokoro, S.; Maruyama, K.; Yoshida, T.; Ishiyama, K.; Kobayashi, M.; et al. Three Arabidopsis SnRK2 Protein Kinases, SRK2D/SnRK2.2, SRK2E/SnRK2.6/OST1 and SRK2I/SnRK2.3, Involved in ABA Signaling are Essential for the Control of Seed Development and Dormancy. *Plant Cell Physiol.* **2009**, *50*, 1345–1363. [CrossRef] [PubMed]

34. Hou, Y.-J.; Zhu, Y.; Wang, P.; Zhao, Y.; Xie, S.; Batelli, G.; Wang, B.; Duan, C.-G.; Wang, X.; Xing, L.; et al. Type One Protein Phosphatase 1 and Its Regulatory Protein Inhibitor 2 Negatively Regulate ABA Signaling. *PLoS Genet.* **2016**, *12*, e1005835. [CrossRef] [PubMed]

35. Fujii, H.; Chinnusamy, V.; Rodrigues, A.; Rubio, S.; Antoni, R.; Park, S.-Y.; Cutler, S.R.; Sheen, J.; Rodriguez, P.L.; Zhu, J.-K. In vitro reconstitution of an abscisic acid signalling pathway. *Nat. Cell Biol.* **2009**, *462*, 660–664. [CrossRef] [PubMed]

36. Lin, Z.; Li, Y.; Zhang, Z.; Liu, X.; Hsu, C.-C.; Du, Y.; Sang, T.; Zhu, C.; Wang, Y.; Satheesh, V.; et al. A RAF-SnRK2 kinase cascade mediates early osmotic stress signaling in higher plants. *Nat. Commun.* **2020**, *11*, 1–10. [CrossRef] [PubMed]

37. Katsuta, S.; Masuda, G.; Bak, H.; Shinozawa, A.; Kamiyama, Y.; Umezawa, T.; Takezawa, D.; Yotsui, I.; Taji, T.; Sakata, Y. Arabidopsis Raf-like kinases act as positive regulators of subclass III SnRK2 in osmostress signaling. *Plant J.* **2020**, *103*, 634–644. [CrossRef] [PubMed]

38. Soma, F.; Takahashi, F.; Suzuki, T.; Shinozaki, K.; Yamaguchi-Shinozaki, K. Plant Raf-like kinases regulate the mRNA population upstream of ABA-unresponsive SnRK2 kinases under drought stress. *Nat. Commun.* **2020**, *11*, 1–12. [CrossRef]

39. Takahashi, Y.; Zhang, J.; Hsu, P.-K.; Ceciliato, P.H.O.; Zhang, L.; Dubeaux, G.; Munemasa, S.; Ge, C.; Zhao, Y.; Hauser, F.; et al. MAP3Kinase-dependent SnRK2-kinase activation is required for abscisic acid signal transduction and rapid osmotic stress response. *Nat. Commun.* **2020**, *11*, 1–12. [CrossRef]

40. Chen, R.; Jiang, H.; Li, L.; Zhai, Q.; Qi, L.; Zhou, W.; Liu, X.; Li, H.; Zheng, W.; Sun, J.; et al. The Arabidopsis Mediator Subunit MED25 Differentially Regulates Jasmonate and Abscisic Acid Signaling through Interacting with the MYC2 and ABI5 Transcription Factors. *Plant Cell* **2012**, *24*, 2898–2916. [CrossRef]

41. Zhao, H.; Nie, K.; Zhou, H.; Yan, X.; Zhan, Q.; Zheng, Y.; Song, C. ABI5 modulates seed germination via feedback regulation of the expression of the PYR/PYL/RCAR ABA receptor genes. *New Phytol.* **2020**. [CrossRef] [PubMed]

42. Finkelstein, R.R.; Lynch, T.J. The Arabidopsis abscisic acid response gene ABI5 encodes a basic leucine zipper transcription factor. *Plant Cell* **2000**, *12*, 599–609. [CrossRef] [PubMed]

43. Laby, R.J.; Kincaid, M.S.; Kim, D.; Gibson, S.I. The Arabidopsis sugar-insensitive mutants sis4 and sis5 are defective in abscisic acid synthesis and response. *Plant J.* **2000**, *23*, 587–596. [CrossRef] [PubMed]

44. Zhu, Y.; Schluttenhoffer, C.M.; Wang, P.; Fu, F.; Thimmapuram, J.; Zhu, J.-K.; Lee, S.Y.; Yun, D.-J.; Mengiste, T. CYCLIN-DEPENDENT KINASE8 Differentially Regulates Plant Immunity to Fungal Pathogens through Kinase-Dependent and -Independent Functions in Arabidopsis. *Plant Cell* **2014**, *26*, 4149–4170. [CrossRef]

45. Elfving, N.; Davoine, C.; Benlloch, R.; Blomberg, J.; Brännström, K.; Müller, D.; Nilsson, A.; Ulfstedt, M.; Ronne, H.; Wingsle, G.; et al. The Arabidopsis thaliana Med25 mediator subunit integrates environmental cues to control plant development. *Proc. Natl. Acad. Sci. USA* **2011**, *108*, 8245–8250. [CrossRef]

46. Kazan, K. The Multitalented MEDIATOR25. *Front. Plant Sci.* **2017**, *8*, 999. [CrossRef] [PubMed]

47. Lai, Z.; Schluttenhofer, C.M.; Bhide, K.; Shreve, J.; Thimmapuram, J.; Lee, S.-Y.; Yun, D.-J.; Mengiste, T. MED18 interaction with distinct transcription factors regulates multiple plant functions. *Nat. Commun.* **2014**, *5*, 3064. [CrossRef]

48. Zhu, Y.; Huang, P.; Guo, P.; Chong, L.; Yu, G.; Sun, X.; Hu, T.; Li, Y.; Hsu, C.; Tang, K.; et al. CDK8 is associated with RAP2.6 and SnRK2.6 and positively modulates abscisic acid signaling and drought response in Arabidopsis. *New Phytol.* **2020**. [CrossRef]

49. Ito, J.; Fukaki, H.; Onoda, M.; Li, L.; Li, C.; Tasaka, M.; Furutani, M. Auxin-dependent compositional change in Mediator in ARF7- and ARF19-mediated transcription. *Proc. Natl. Acad. Sci. USA* **2016**, *113*, 6562–6567. [CrossRef]

50. Harper, T.M.; Taatjes, D.J. The complex structure and function of Mediator. *J. Biol. Chem.* **2017**, *293*, 13778–13785. [CrossRef]

51. Chen, J.; Mohan, R.; Zhang, Y.; Li, M.; Chen, H.; Palmer, I.A.; Chang, M.; Qi, G.; Spoel, S.H.; Mengiste, T.; et al. NPR1 Promotes Its Own and Target Gene Expression in Plant Defense by Recruiting CDK8. *Plant Physiol.* **2019**, *181*, 289–304. [CrossRef] [PubMed]

52. Huang, J.; Sun, Y.; Orduna, A.R.; Jetter, R.; Li, X. The Mediator kinase module serves as a positive regulator of salicylic acid accumulation and systemic acquired resistance. *Plant J.* **2019**, *98*, 842–852. [CrossRef] [PubMed]

53. Gong, Z.; Xiong, L.; Shi, H.; Yang, S.; Herrera-Estrella, L.R.; Xu, G.; Chao, D.-Y.; Li, J.; Wang, P.-Y.; Qin, F.; et al. Plant abiotic stress response and nutrient use efficiency. *Sci. China Life Sci.* **2020**, *63*, 635–674. [CrossRef] [PubMed]

54. Knight, H.; Veale, E.L.; Warren, G.J.; Knight, M.R. The sfr6 Mutation in Arabidopsis Suppresses Low-Temperature Induction of Genes Dependent on the CRT/DRE Sequence Motif. *Plant Cell* **1999**, *11*, 875. [CrossRef] [PubMed]

55. Boyce, J.M.; Knight, H.; Deyholos, M.; Openshaw, M.R.; Galbraith, D.W.; Warren, G.; Knight, M.R. The sfr6 mutant of Arabidopsis is defective in transcriptional activation via CBF/DREB1 and DREB2 and shows sensitivity to osmotic stress. *Plant J.* **2003**, *34*, 395–406. [CrossRef]

56. Jaglo-Ottosen, K.R. Arabidopsis CBF1 Overexpression Induces COR Genes and Enhances Freezing Tolerance. *Science* **1998**, *280*, 104–106. [CrossRef]

57. Knight, H.; Mugford, S.G.; Ülker, B.; Gao, D.; Thorlby, G.; Knight, M.R. Identification of SFR6, a key component in cold acclimation acting post-translationally on CBF function. *Plant J.* **2009**, *58*, 97–108. [CrossRef]

58. Hemsley, P.A.; Hurst, C.H.; Kaliyadasa, E.; Lamb, R.; Knight, M.R.; De Cothi, E.A.; Steele, J.F.; Knight, H. The Arabidopsis mediator complex subunits MED16, MED14, and MED2 regulate mediator and RNA polymerase II recruitment to CBF-responsive cold-regulated genes. *Plant Cell* **2014**, *26*, 465–484. [CrossRef]

59. Yang, Y.; Ou, B.; Zhang, J.; Si, W.; Gu, H.; Qin, G.; Qu, L.J. The Arabidopsis Mediator subunit MED16 regulates iron homeostasis by associating with EIN3/EIL1 through subunit MED25. *Plant J. Cell Mol. Biol.* **2014**, *77*, 838–851. [CrossRef]

60. Gong, Z. Genes That Are Uniquely Stress Regulated in Salt Overly Sensitive (sos) Mutants. *Plant Physiol.* **2001**, *126*, 363–375. [CrossRef]

61. Zhu, J.K. Salt and drought stress signal transduction in plants. *Annu. Rev. Plant Biol.* **2002**, *53*, 247–273. [CrossRef] [PubMed]

62. Zhu, Y.; Wang, B.; Tang, K.; Hsu, C.-C.; Xie, S.; Du, H.; Yang, Y.; Tao, W.A.; Zhu, J.-K. An Arabidopsis Nucleoporin NUP85 modulates plant responses to ABA and salt stress. *PLoS Genet.* **2017**, *13*, e1007124. [CrossRef]

63. Crawford, T.; Karamat, F.; Lehotai, N.; Rentoft, M.; Blomberg, J.; Strand, Å.; Björklund, S. Specific functions for Mediator complex subunits from different modules in the transcriptional response of Arabidopsis thaliana to abiotic stress. *Sci. Rep.* **2020**, *10*, 1–18. [CrossRef]

64. Morimoto, K.; Mizoi, J.; Qin, F.; Kim, J.-S.; Sato, H.; Osakabe, Y.; Shinozaki, K.; Yamaguchi-Shinozaki, K. Stabilization of Arabidopsis DREB2A Is Required but Not Sufficient for the Induction of Target Genes under Conditions of Stress. *PLoS ONE* **2013**, *8*, e80457. [CrossRef] [PubMed]

65. Qin, F.; Sakuma, Y.; Tran, L.-S.P.; Maruyama, K.; Kidokoro, S.; Fujita, Y.; Fujita, M.; Umezawa, T.; Sawano, Y.; Miyazono, K.-I.; et al. Arabidopsis DREB2A-Interacting Proteins Function as RING E3 Ligases and Negatively Regulate Plant Drought Stress–Responsive Gene Expression. *Plant Cell* **2008**, *20*, 1693–1707. [CrossRef]

66. Zhu, Q.; Zhang, J.; Gao, X.; Tong, J.; Xiao, L.; Li, W.; Zhang, H. The Arabidopsis AP2/ERF transcription factor RAP2.6 participates in ABA, salt and osmotic stress responses. *Gene* **2010**, *457*, 1–12. [CrossRef]

67. Djemal, R.; Khoudi, H. Isolation and molecular characterization of a novel WIN1/SHN1 ethylene-responsive transcription factor TdSHN1 from durum wheat (Triticum turgidum. L. subsp. durum). *Protoplasma* **2015**, *252*, 1461–1473. [CrossRef] [PubMed]

68. Djemal, R.; Mila, I.; Bouzayen, M.; Pirrello, J.; Khoudi, H. Molecular cloning and characterization of novel WIN1/SHN1 ethylene responsive transcription factor HvSHN1 in barley (Hordeum vulgare L.). *J. Plant Physiol.* **2018**, *228*, 39–46. [CrossRef] [PubMed]

Publisher's Note: MDPI stays neutral with regard to jurisdictional claims in published maps and institutional affiliations.

International Journal of
Molecular Sciences

Article

Kaolin Reduces ABA Biosynthesis through the Inhibition of Neoxanthin Synthesis in Grapevines under Water Deficit

Tommaso Frioni [1], Sergio Tombesi [1,*], Paolo Sabbatini [2], Cecilia Squeri [1],
Nieves Lavado Rodas [3], Alberto Palliotti [4] and Stefano Poni [1]

[1] Department of Sustainable Crop Production, Università Cattolica del Sacro Cuore, Via Emilia Parmense 84, 29122 Piacenza, Italy; tommaso.frioni@unicatt.it (T.F.); cecilia.squeri@unicatt.it (C.S.); stefano.poni@unicatt.it (S.P.)
[2] Department of Horticulture, Michigan State University, 1066 Bogue Street, East Lansing, MI 48824, USA; sabbatin@msu.edu
[3] CICYTEX (Junta de Extremadura), Finca La Orden, Ctra. A-V, km 372, Guadajira, 06187 Badajoz, Spain; nieves.lavado@juntaex.es
[4] Department of Agricultural, Food and Environmental Sciences, Università degli Studi di Perugia, Borgo XX Giugno 74, 06121 Perugia, Italy; alberto.palliotti@unipg.it
* Correspondence: sergio.tombesi@unicatt.it; Tel.: +390523599221

Received: 4 June 2020; Accepted: 10 July 2020; Published: 13 July 2020

Abstract: In many viticulture regions, multiple summer stresses are occurring with increased frequency and severity because of warming trends. Kaolin-based particle film technology is a technique that can mitigate the negative effects of intense and/or prolonged drought on grapevine physiology. Although a primary mechanism of action of kaolin is the increase of radiation reflection, some indirect effects are the protection of canopy functionality and faster stress recovery by abscisic acid (ABA) regulation. The physiological mechanism underlying the kaolin regulation of canopy functionality under water deficit is still poorly understood. In a dry-down experiment carried out on grapevines, at the peak of stress and when control vines zeroed whole-canopy net CO_2 exchange rates/leaf area (NCER/LA), kaolin-treated vines maintained positive NCER/LA (~2 μmol m^{-2} s^{-1}) and canopy transpiration (E) (0.57 μmol m^{-2} s^{-1}). Kaolin-coated leaves had a higher violaxanthin (Vx) + antheraxanthin (Ax) + zeaxanthin (Zx) pool and a significantly lower neoxanthin (Nx) content (VAZ) when water deficit became severe. At the peak of water shortage, leaf ABA suddenly increased by 4-fold in control vines, whereas in kaolin-coated leaves the variation of ABA content was limited. Overall, kaolin prevented the biosynthesis of ABA by avoiding the deviation of the VAZ epoxidation/de-epoxidation cycle into the ABA precursor (i.e., Nx) biosynthetic direction. The preservation of the active VAZ cycle and transpiration led to an improved dissipation of exceeding electrons, explaining the higher resilience of canopy functionality expressed by canopies sprayed by kaolin. These results point out the interaction of kaolin with the regulation of the VAZ cycle and the active mechanism of stomatal conductance regulation.

Keywords: particle film technology; xanthophylls; VAZ cycle; drought; *Vitis vinifera* L.; abscisic acid

1. Introduction

Global warming is rapidly changing worldwide agriculture. Growers are facing general warming trends that intensify extreme events and compromise yield and fruit quality [1–3]. Viticulture is one of the most relevant crops in warm and temperate regions and, in Mediterranean wine districts, multiple summer stresses (i.e., the concurrence of prolonged drought, high air temperature and excessive light radiation) are the main causes of vineyard impairments related to climate change [3,4]. The first consequence of multiple summer stresses is the reduction of carbon assimilation and transpiration (E), resulting in the loss of yield and fruit quality, according to the severity and duration of limiting conditions [4–6].

Kaolin (aluminium silicate) is a natural clay used for mitigating the negative impact of extreme temperatures on leaves and fruits [7–9]. Under non-limiting conditions, mild or no effects on assimilation and transpiration have been reported [10–15]. Under water shortage, kaolin is able to maintain better assimilation and transpiration rates, to avoid photosystem II efficiency loss and to prevent leaf photoinhibition and abscission [9,12–15]. Its mechanism of action is primarily related to the increase of light reflection that reduces the radiation absorbed by the leaf [7,8,16].

Leaf transpiration is controlled by stomata, which under water deficit are subjected to active and passive regulation. Passive regulation relies on hydraulic mechanisms mediated by environmental conditions and xylem water potential [17–19], whereas active mechanisms consist of biochemical signalling of abscisic acid (ABA), the accumulation of which affects stomatal closure in leaves [20–22].

ABA is biosynthesised in roots and leaves [23], although leaf biosynthesis seems the predominant one and the one responsible for the accumulation of ABA in leaves [24]. ABA biosynthesis proceeds from the degradation of carotenoids involved in the xanthophylls cycle (VAZ cycle) [23]. During a warm day, violaxanthin (Vx) is de-epoxidated to zeaxanthin (Zx) via the formation of antheraxanthin (Ax) [25]. The activity of Vx de-epoxide is regulated by chloroplast stroma acidification caused by the accumulation of electrons not used for the biochemical reactions of photosynthesis. Thus, the reduction of photosynthetic activity triggers the de-epoxidation of Vx, contributing to the active energy dissipation even when thermoregulation through transpiration is missing or reduced [25]. Zx is then epoxidated by the Zx epoxidase enzyme (ZEP) during a dark period, being ZEP inhibited by light [26]. On the other hand, Zx is the precursor of neoxanthin (Nx), which is formed by the activity of Nx synthase [27]. Nx is cleaved to xanthoxin and ABA-aldehyde, the reaction of which is catalysed by the 9-cis-epoxycarotenoid dioxygenase enzyme (NCED) [23]. ABA biosynthesis is induced by leaf dehydration and by the decline of cell volume [22,24]. In the ABA biosynthetic pathway, *NCED* is the gene involved in the promotion of leaf dehydration due to the increase of air-to-leaf vapor pressure deficit (VPD) [28].

Recently, in mature grapevines grown in the field, Dinis et al. [29] showed a kaolin-induced reduction of leaf ABA associated with better leaf stomatal conductance than untreated vines at leaf water potential lower than -1 MPa. Brito et al. [16] reported similar responses in olive trees.

However, the causes and the consequences of the kaolin-induced modulation of ABA biosynthesis were never investigated and the dynamic interactions between kaolin and leaf Vx, Ax, Zx and Nx under progressive water deficit are currently unknown. Since ABA biosynthesis is connected to the VAZ cycle, our hypothesis was that a lower leaf ABA in kaolin-coated leaves could be associated with a different tuning of the de-epoxidation/epoxidation state under water deficit. Therefore, the aim of the present work was to determine:

1. if Kaolin had an effect on ABA biosynthetic pathway,
2. if the eventual difference in ABA accumulation was related to possible bottlenecks on the carotenoid biosynthetic pathway that leads to ABA biosynthesis in leaves.

2. Results

2.1. Leaf Physical Properties and Vine Physiology

Kaolin coating increased leaf light reflection, increasing from 10% of the total incident radiation recorded in control vines to the 15% recorded in sprayed leaves (Table 1). Conversely, leaf-transmitted light was reduced to 6.9% of total photosynthetically active radiation (PAR) in kaolin-treated leaves vs. 8.0% found in control vines.

Table 1. Photosynthetically active radiation (PAR) reflected and transmitted at 1:00 p.m. by kaolin.

	Reflected PAR			**Transmitted PAR**		
	(% of Total PAR)			(% of Total PAR)		
Control	10.10%	±	0.88 b	8.30%	±	0.02 a
Kaolin	15.14%	±	0.45 a	6.86%	±	0.41 b

Different letters mean significant difference per $p < 0.05$ (t-test).

Leaf temperature (T_{leaf}) was similar in the two treatments except at Day Of the Year (DOY) 215 to 216 and 217 (the 3 last days of water deficit) when kaolin-coated leaves were 2.4 °C cooler than control leaves (Figure 1A). There were no differences upon re-watering (after DOY 217). Kaolin did not affect stem water potential at midday (Ψ_{md}), which decreased in both treatments from initial −0.6 MPa to −1.7 MPa on DOY 217 following the same pattern (Figure 1B). After re-watering, Ψ_{md} was restored to pre-stress value ranges. In both treatments, transpiration rate/leaf area (E/LA) decreased significantly from DOY 208 (Figure 1C). No differences between treatments were found until DOY 215, when control vines had significantly lower E/LA than kaolin vines, reaching on DOY 217 a value of 0.37 and 0.57 mmol m^{-2} s^{-1}, respectively. After re-watering, no differences were found again between treatments. Net CO_2 exchange rates/leaf area (NCER/LA) (Figure 1D) followed a similar pattern—kaolin vines maintained higher assimilation rates from DOY 215 to DOY 217 (~2 µmol m^{-2} s^{-1}), whereas control vines zeroed their NCER/LA. Upon re-watering (DOY 219), kaolin exhibited higher NCER/LA than control (+46%), but this effect vanished at DOY 228, 10 days after the restoration of full water supply.

No difference between treatments was found in the response of E/LA to varying Ψ_{md} after kaolin coating (Figure 2). Independently of the treatment, E/LA was positively correlated to Ψ_{md} ($y = 1.17x + 2.03$, $R^2 = 0.52$, $p < 0.05$).

2.2. Xanthophylls and ABA Concentration

In kaolin leaves, total xanthophylls (Vx + Ax + Zx) concentration was significantly higher than control leaves from DOY 208 to DOY 219, the first day after re-watering. In this time span, the difference between treatments ranged between 107 and 367 µg g dw^{-1} (Figure 3A). Ten days after re-watering, these differences between kaolin and control disappeared. The de-epoxidation state set at about 0.4 in both treatments between DOY 205 and 207 (Figure 3B). The proportion of de-epoxidated xanthophylls was consistently higher in kaolin vs. control at DOY 211, 212 and 214, while at DOY 216 de-epoxidated xanthophylls were consistently higher in control than in the kaolin treatment.

Leaf Nx concentration was similar between treatments during the first part of the experiment (Figure 3C). However, starting from DOY 212, there was a clear course indicating higher Nx in control leaves. The difference between kaolin and control peaked on DOY 216, when Nx in control leaves was more than 3-fold that in kaolin leaves. Leaf ABA ranged between 0.5 and 2 ng g dw^{-1} in both treatments until DOY 215 (Figure 3D). On DOY 216 and 217, leaf ABA in control vines suddenly peaked up to 8.1 ng g dw^{-1}, whereas in kaolin vines leaf ABA mildly increased up to 3.2 ng g dw^{-1}. Right after re-watering, no difference between treatments was found; however, 10 days after the restoration of full water supply, control leaves showed again a significantly higher leaf ABA concentration (1.3 µg g dw^{-1}).

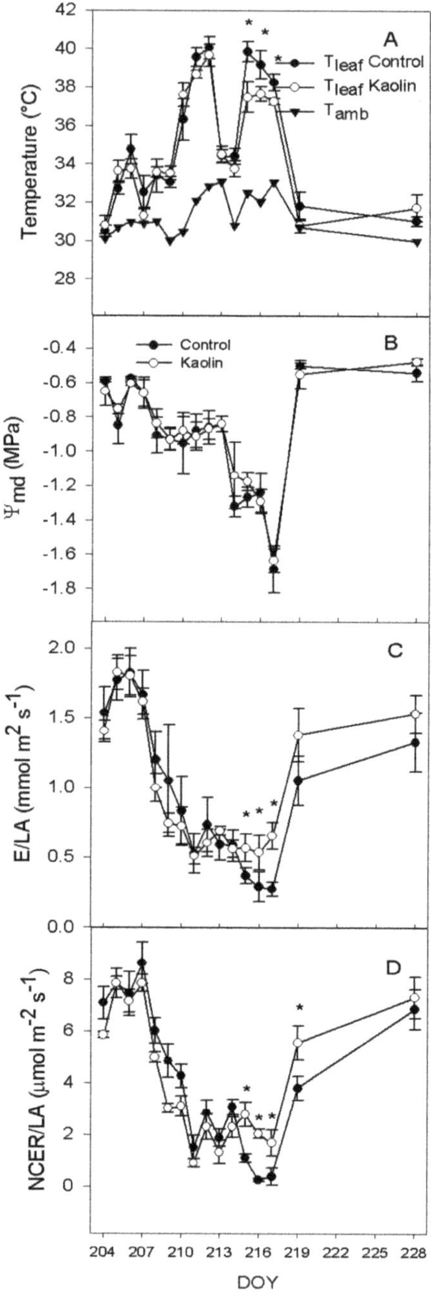

Figure 1. (**A**)Trends for air (T_{amb}) and leaf (T_{leaf}) temperature; (**B**)midday stem water potential (Ψ_{MD}); (**C**) whole-canopy transpiration (E/LA) and (**D**) specific whole-canopy net CO_2 exchange rate/leaf area (NCER/LA), according to a progressive water shortage (DOY 209–217) and subsequent re-watering (at DOY 218), in vines subjected to the kaolin treatment and in controls. Bars represents standard error (SE), $n = 3$. Asterisks indicate dates within which differences among treatment were significant ($p < 0.05$). DOY: day of the year.

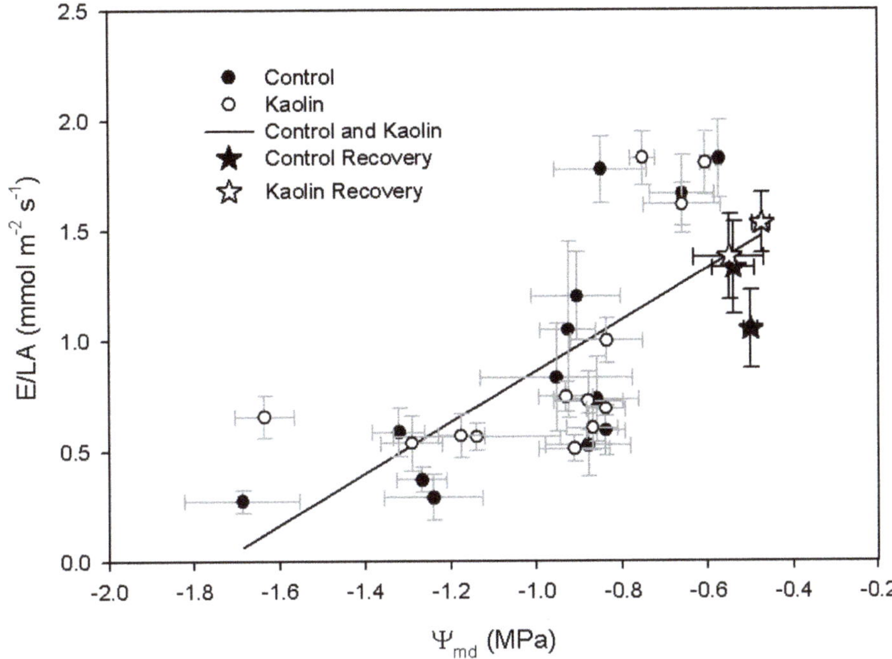

Figure 2. Correlation between whole-canopy transpiration rate/leaf area (E/LA) and midday stem water potential (Ψ_{MD}).

Vx leaf concentration at 4:00 a.m. decreased in both treatments until DOY 213 (Figure 4A). From DOY 214 to DOY 216, kaolin vines had a higher leaf Vx concentration than control vines (53 µg g dw^{-1}, if averaged over the three days). Although no difference was found at the peak of water deficit, higher leaf Vx concentration in kaolin vines was found again upon re-watering (90 µg g dw^{-1}). The night recovery of Vx was variable during the experiment, yet no differences were found between treatments throughout the experiment (Figure 4B). From DOY 205 to DOY 209, kaolin did not affect leaf Zx concentration (Figure 4C). On DOY 210, Zx concentration was higher in kaolin leaves (184 ± 41 µg g dw^{-1} vs. 72 ± 36 µg g dw^{-1} found in control). Difference between treatments peaked on DOY 212 (369 µg g dw^{-1} in kaolin) and vanished once vines were re-watered.

The correlation between E/LA and ABA was described by an exponential decay function ($y = 0.28 + 2.02e^{-0.75x}$; $R^2 = 0.73$, $p < 0.0001$), with no difference due to the treatments, although the range of ABA covered by kaolin vines was lower than those covered by control ones (Figure 5). In kaolin-treated vines, a significant positive linear correlation ($y = -7.16 + 0.25x$; $R^2 = 0.61$, $p < 0.005$) was found for leaf ABA vs. T$_{leaf}$ (Figure 6). On the contrary, in control vines leaf ABA was not correlated to varying T$_{leaf}$.

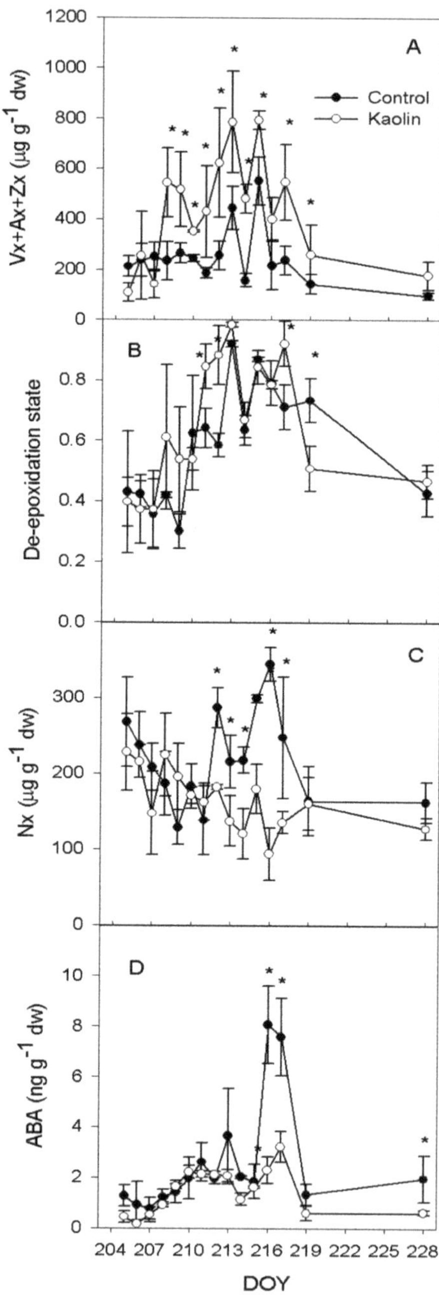

Figure 3. (**A**) Trends for leaf violaxanthin (Vx) + antheraxanthin (Ax) + zeaxanthin (Zx) content, (**B**) de-epoxidation state, (**C**) neoxanthin (Nx) content and (**D**) abscisic acid (ABA) concentration, according to a progressive water shortage (DOY 209–217) and subsequent re-watering (at DOY 218), in vines subjected to the kaolin treatment and in controls. Bars represents standard error, $n = 3$. Asterisks indicate dates within which differences among treatment were significant ($p < 0.05$). DOY: day of the year.

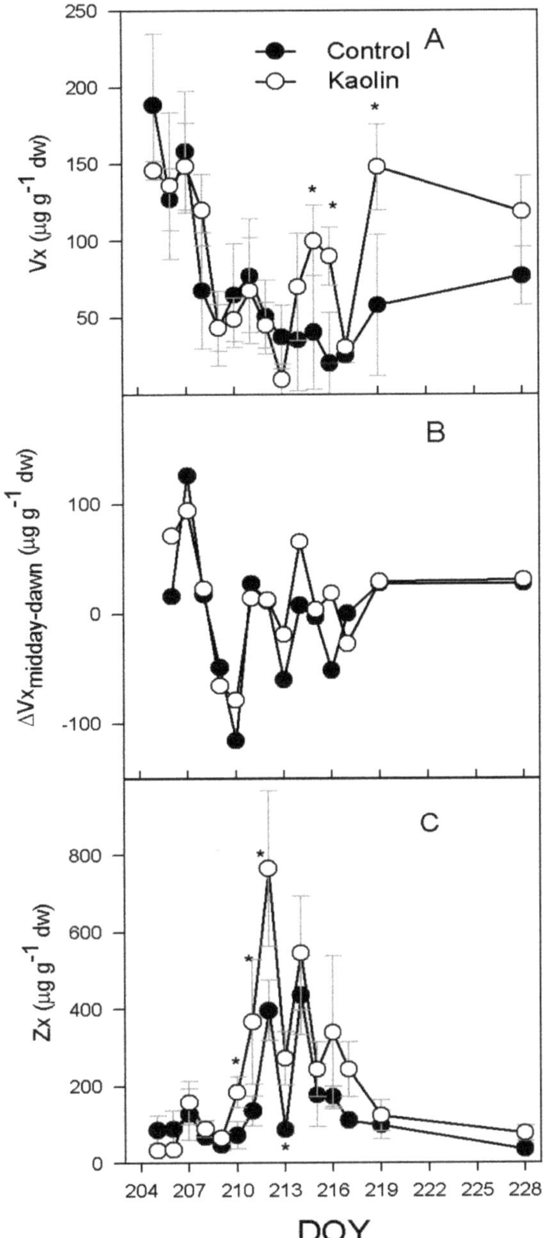

Figure 4. (**A**) Course of leaf violaxanthin (Vx) content at dawn; (**B**) midday to dawn Vx differences and (**C**) zeaxanthin (Zx) content at midday over the experiment, according to a progressive water shortage (DOY 209–217) and subsequent re-watering (at DOY 218), in vines subjected to the kaolin treatment and in controls. Bars represents standard error, $n = 3$. Asterisks indicate dates within which differences among treatment were significant ($p < 0.05$). DOY: day of the year.

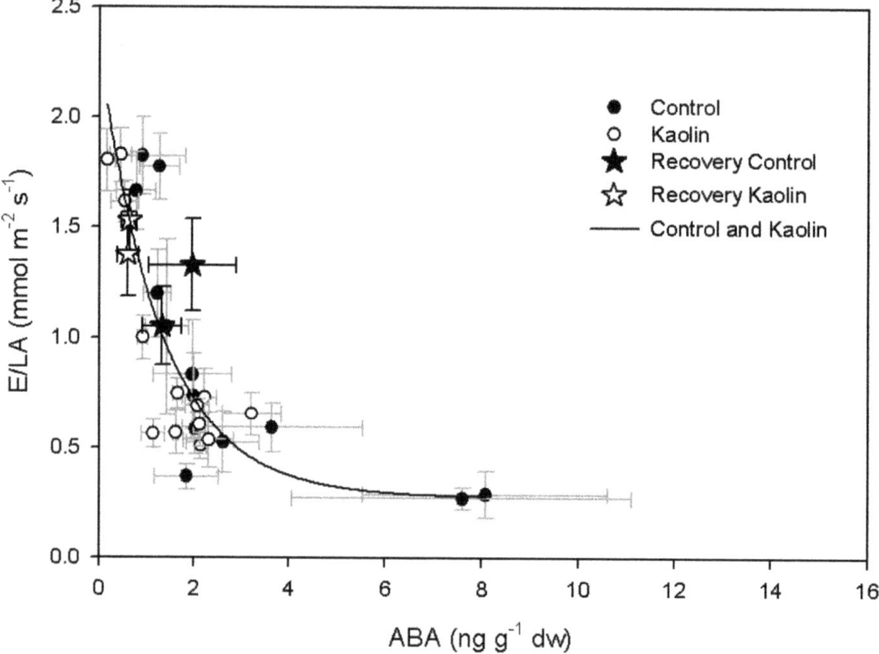

Figure 5. Correlation between whole-canopy transpiration rate/leaf area (E/LA) and leaf abscisic acid (ABA) content.

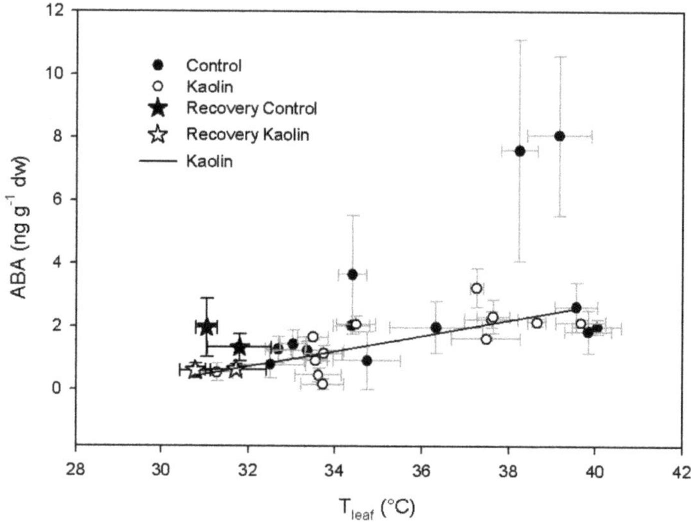

Figure 6. Correlation between leaf abscisic acid (ABA) content and leaf temperature (T_{leaf}).

3. Discussion

The regulation of transpiration occurs through the response of the whole vine physiology to multiple environmental signals and biochemical regulation. In this experiment, we limited the light absorbed by leaves by spraying kaolin, a natural clay inducing light reflection in the visible, infra-red and ultra-violet wavebands [8,30]. Kaolin-based formulates are widely used in many crops to reduce leaf and fruit sunburn damages [7,8,11,13,31–33]. In our experiment, kaolin was effective at increasing single leaf reflected light and at decreasing transmitted light (Table 1), in agreement with Rosati et al. [10] and Steiman et al. [30].

In our experiment, when vines were exposed to reduced water supply, Ψ_{md} decreased and leaf transpiration slowed down along with photosynthesis (Figure 1). Overall, the kaolin treatment had a moderate effect on transpiration rate, mainly exerted under severe water deficit, resulting in a reduction of T_{leaf}. The difference of T_{leaf} between treatments appeared to be related to the larger transpiration, rather to the reduction of absorbed light; prior to water stress (WS) imposition, T_{leaf} was similar in the two treatments. On the other hand, during water shortage, T_{leaf} was significantly lower in kaolin in comparison with control.

Transpiration is regulated by stomata—stomatal conductance is influenced by environmental stimuli, such as light and leaf-to-air vapour pressure deficit, and by physiological regulation mechanisms that are usually divided into active, mainly ABA-mediated, and passive, mainly related to water potential [18–20]. In our experiment, we did not observe, between treatments, significant variation of stomata response to the decrease of leaf water potential and to the increase of leaf ABA (Figures 2 and 5). The correlation between transpiration rate and leaf water potential was linear; instead, in the plot of E/LA vs. ABA, data were divided in three clusters as follows: data recorded before water deficit onset, when E/LA was above 1.5 mmol m^{-2} s^{-1}; data recorded during reduced water supply; and the two points of the control vines recorded at the end of the water deficit period. During water recovery, E values were similar to those recorded during water stress. These data are consistent with previous studies that reported a decrease of transpiration as related to leaf water potential during the early stages of water shortage, followed by a significant increase of ABA after stomata closure [34,35]. However, this was not the case in kaolin vines, where ABA content did not significantly increase during the final stages of water deficit, as was instead the case in control vines. Overall, these data are in agreement with Dinis et al. [29], who reported a linear correlation between stomatal conductance (g_s) and leaf ABA of field-grown grapevines, even though in their work kaolin-coated leaves had a looser response of g_s to leaf ABA concentrations, in comparison with control. Despite the evidence of a lower ABA content in kaolin-treated leaves, its cause is still unclear.

The ABA precursor is neoxanthin (Nx), which is formed from zeaxanthin (Zx), after the de-epoxidation of violaxanthin (Vx), via the intermediate antheraxanthin (Ax) formation. In our experiment, the amount of Vx + Ax + Zx, representing the molecules involved in the xanthophyll (VAZ) cycle, increased in the kaolin treatment, in comparison with control. Light plays an important role in the activation of the VAZ cycle—the regulation of violaxanthin de-epoxidase enzyme (VDE) is pH dependent and it is activated by the acidification of the thylakoid lumen when photosynthetic electron transport exceeds the capacity of assimilatory reactions [25]. In our experiment, the reduction of water supply promoted the de-epoxidation state of xanthophylls, while in control there was an increase of Nx and, in the kaolin treatment, the Nx content was almost constant during all the experiment (Figure 3C). This suggests that, theoretically, the kaolin treatment reduced the conversion of Zx into Nx by stimulating the activity of ZEP or by reducing the activity of neoxanthin synthase (NxS). ZEP is inhibited by light, and the epoxidation of Zx to Vx mainly occurs at night [26]. In our experiment, the increase of Vx after the night recovery ($\Delta Vx_{midday-dawn}$) was similar (Figure 4B), although, in kaolin, the larger pool of Vx + Ax + Zx during the water stress period (Figure 3A) led to a significantly higher content, in comparison with control, at DOY 215 and 216 (Figure 4A). These data suggest a similar activity of ZEP in the two treatments. On the other hand, from DOY 210 to 213 and from DOY 216 to 217, there was a significantly larger content of Zx in kaolin, in comparison with control, whereas Nx was lower. Therefore, the conversion of Zx into Nx was likely limited in kaolin vines. Nx is the precursor of ABA, which followed the same dynamic of Nx in both treatments during the experiment, while in kaolin ABA had a modest increase during the experiment and, in control, ABA increased almost 4-fold. This could explain the further decrease of transpiration observed in control vines vs. kaolin ones from

DOY 215 to 217. In kaolin, the lack of Nx increase limited the ABA leaf content when water deficit was more severe and leaf physiology more affected by leaf dehydration. Interestingly, the contrasting behaviour in the two treatments regarding Nx and ABA biosynthesis was not related to Ψ_{md}, which was similar across the two treatments (Figure 1B), neither to T_{leaf}, since ABA content was correlated to leaf temperature only in the kaolin treatment (Figure 6). Our results suggest that the higher ABA biosynthetic rates were primarily driven by the prolonged reduction of E, resulting in an increase of T_{leaf} concurrent to the increase of leaf ABA.

These results contribute to explaining the mechanisms involved in the kaolin-induced protection of canopy functionality, that is, in reflecting radiation and preserving thermoregulation, kaolin maintains a viable leaf VAZ cycle running. This, in turn, avoids the energy excess, potentially leading to damage to photosystems. An active VAZ cycle prevents the onset of ABA biosynthesis by reducing the accumulation of its precursors.

4. Materials and Methods

4.1. Plant Material and Treatment Layout

The experiment was carried out in 2018 in an outdoor area in Piacenza (44°55′ N, 9°44′ E), Italy, close to the Agriculture faculty, on 6 five-year-old cultivar Sangiovese (*Vitis vinifera* L.) vines (clone R10, grafted on SO4 rootstock) grown in 55l pots. The set of plants was similar to that used in Frioni et al. [15]. In this experiment, six vines were arranged along a single row having a 35° NE-SW orientation. Vines were cane pruned, trained to vertically shoot-positioned (VSP) trellis. Horizontal cane was 1 m long accommodating 9 nodes and it was raised 90 cm from the ground. The pots were filled with a mixture of loamy soil and peat (80:20 by volume, respectively) and kept well watered until the beginning of the water deficit. Pots were painted white before the trial started, to limit radiation-induced overheating. Each vine was fertilized twice (i.e., one week before and two weeks after bud-break) with 4 g of Greenplant 15 (N) + 5 (P_2O_5) + 25 (K_2O) +2 (MgO) + micro (Green Has Italia, Cuneo, Italy). The six vines were then randomly divided into two treatments as follows: three vines were sprayed on 23 July (DOY 204) at 9:00 a.m. with a formulation of 100% aluminium silicate (Baïkal, Agrisynergie, Périgueux, France) diluted in water at 3% concentration (kaolin); the remaining three vines were assigned to the untreated control (control). The kaolin solution was carefully sprayed on both canopy sides with a shoulder pump.

For the dry-down setup, the same protocol described in Frioni et al. [15] was used. All the vines were kept well watered until DOY 208 (27 July, phenological stage BBCH77 according to Lorenz et al. [36]) by supplying a daily amount of 3600 ± 424 ml per vine, representing 110% actual canopy transpiration (E) concurrently measured by the whole-canopy system described hereafter over DOY 200–207. Irrigation was performed through an automated water supply system described by Poni et al. [37]. Starting on DOY 209 (28 July), a constant water deficit was imposed on all the vines by programming the water supply system to deliver daily to each vine only 70% of whole-canopy potential transpiration, calculated on the basis of the data collected prior to the water deficit imposition (DOY 200–207), until the achievement of severe water deficit conditions. Re-watering was performed on DOY 218 (6 August), restoring full water supply to all vines until dismantling of the chambers. During water shortage, each pot surface was covered with a plastic sheet to prevent infiltration of rain water and to minimize losses due to soil evaporation.

4.2. Whole-Canopy Gas Exchange

Whole-canopy net CO_2 exchange rate/leaf area (NCER/LA) measurements were taken using the multi-chamber system reported in Poni et al. [37] with the configuration described in Frioni et al. [15]. To warrant unbiased comparison vs. canopy development, leaf area (LA) per vine was estimated as described in Gatti et al. [38] and NCER/LA (μmol CO^2/m^2s) computed accordingly. Since vines assigned to the two treatments had the same shoot number inside (~8) and outside (1) the chambers and, additionally, shoot growth along the cane was very uniform, measured E and NCER were estimated to be ~91% of total vine actual transpiration and assimilation.

The chambers were set up on each vine and continuously operated 24 h from DOY 200 (19 July, four days prior to kaolin sprays and eight days prior to the beginning of reduced water supply) until DOY 229 (17 August, 10 days after re-watering of WS plants). Daily data were screened to consider only data recorded between 11:00 a.m. and 3:00 p.m. (mean PAR > 1000 μmol photons $m^{-2}s^{-1}$) and in the absence of rain or unfavourable weather conditions. Ambient (inlet) air temperature was measured by shielded 1/0.2 mm diameter PFA-Teflon insulated type-T thermocouples (Omega Eng. INC, Stamford, CT, USA) and direct and diffuse radiation were measured with a BF2 sunshine sensor (Delta-T Devices, Ltd, Cambridge, UK) placed horizontally on top of a support stake next to the chambers enclosing the canopies. Ambient (inlet) relative humidity (RH) at each chamber's outlet was measured by an HIH-4000 humidity sensor (Honeywell, Freeport, IL, USA) mounted upstream of the EGM-4 (PP system, Amesbury, MA, USA). Chambers were dismantled on 18 August (DOY 230).

4.3. Leaf Water Status, Temperature and Light Reflectance

Progression of water deficit was monitored by measuring midday stem water potential (Ψ_{MD}). Ψ_{MD} was measured daily at 1:00 p.m. on days with clear sky on one leaf per vine. Measures were taken using a Scholander pressure chamber (3500 Model, Soilmosture Equip. Corp., Santa Barbara, CA, USA). Leaves were sampled from the shoots contained in the chamber using a custom-built zip-lock lateral access to the chamber.

On the same days, mean temperature (T_{leaf}) was measured on a mature well-exposed leaf per vine inserted on the shoot kept outside of the chambers. On each date, one frontal thermal image per leaf was taken under full sunlight conditions at ~50 cm distance from the leaf itself, using the FLIR i60 infra-red thermal imaging camera (FLIR Systems Inc., Wilsonville, OR, USA). Thermal image analysis was carried out with the FLIR Tools software (FLIR Systems Inc).

On DOY 212 (31 July), the PAR transmitted through and reflected from the leaves was measured using a Sun System PAR-meter with external sensor (Sun System, Vancouver, WA, USA) as described by Taylor [39]. For the reflected PAR, the light sensor was downward oriented about 50 cm above a fully exposed leaf; for the transmitted PAR, the light sensor was upward oriented, holding it 50 cm below the leaf, on its projected shade. Measures were taken on DOY 212 (31 July) at 1:00 p.m. on two leaves per vine.

4.4. Xanthophylls and Abscisic Acid Determination

The leaf concentration of abscisic acid (ABA), violaxanthin (Vx), antheraxanthin (Ax), zeaxanthin (Zx) and neoxanthin (Nx) was determined on fully mature leaves sampled at 2:00 p.m. from the beginning to the last day of the experiment. Additionally, Vx was also analysed in leaves sampled at dawn (4:00 a.m.). Three leaves per treatment were cut, immediately wrapped in aluminium foil and dipped into liquid N_2. In a preliminary experiment, we sampled 10 leaves and, after cutting them in two halves, we washed one half and we proceeded to Vx, Ax, Zx, Nx and ABA determination using the same procedure described below. No significant differences between washed and unwashed samples were found ($p = 0.05$). The samples were stored at $-80\ °C$ and then lyophilised. Lyophilised material was weighed (dry weight) and ground.

Vx, Ax, Zx and Nx extraction was carried out following the method by Yuan and Qian [40] with some modifications. First, 100 mg of lyophilised leaf were spiked with 100 μl of internal standard (100 mg/L β-apo-8'-carotenal). Extraction was carried out with 1 mL of ethyl acetate containing 0.1% BHT, kept in agitation for half an hour. After centrifugation at 1500 rpm for 5 min, the resulting upper layer was collected, whereas the lower layer was extracted again with 1 mL of ethyl acetate containing 0.1% BHT. The combined upper layer extracts were dried by evaporation at 30 °C under vacuum. The residue was resuspended in 1 mL of acetone containing 0.1% BHT and centrifuged at 11,000 rpm for 5 min. The clear extract was injected onto the HPLC. Each sample was extracted in triplicate.

Sample handling, homogenization and extraction were carried out under reduced light and kept cold to minimize light-induced isomerization and oxidation of carotenoids. The concentration of carotenoids was determined on an Agilent 1260 Infinity HPLC (Santa Clara, CA, USA). A Luna 3 μm C8 (2) 100A, 100×4.6 mm column (Phenomenex, Castel Maggiore, Italy) was used with the setting temperature at 60 °C. The eluents were methanol and 1 M ammonium acetate (70:30) (solvent A) and 100% methanol (solvent B). Total flow rate was 1 mL/min. A binary gradient system was employed passing from 0 min (95% A 5% B) to 60 min (5% A; 95% B). Sample injection volume was 5 μL, and absorbance at 450 nm was used for quantification. β-Carotene was identified by comparison with retention time and UV spectra of commercial β-carotene standard (95% purity) (Sigma-Aldrich, St. Louis, MO, USA). Identification of the other carotenoids was performed by comparing retention time and UV–visible photodiode array spectra with authentic standards. Vx, Ax, Zx and Nx standards were purchased from CaroteNature (Münsingen, Switzerland). All the compounds were run in triplicate and calculated as β-carotene equivalent. De-epoxidation state was calculated as $(Zx + 0.5Ax)/(Vx + Ax + Zx)$. Vx variation between midday and the next dawn ($\Delta Vx_{midday\text{-}dawn}$) was calculated as the difference between the mean Vx content at midday and the mean Vx content at dawn per each treatment.

ABA was extracted following the procedure described by Vilarò et al. with some modifications [41]. Leaf material (0.1 g) was extracted with 10 mL of methanol/water (1:1 v/v, pH = 3 with formic acid) and homogenised for 3 min with an Ultra-Turrax homogenizer (IKA-Werke GmbH & Co., Staufen, Germany). After centrifugation (5000 rpm for 6 min), the supernatant was filtered through a paper filter and the same procedure was repeated for the remaining pellet. The collected filtrates were extracted twice with dichloromethane (15 mL) and the organic phase evaporated under vacuum. The residue was dissolved to a 100 μL acetone and 250 μL water/acetonitrile (70:30 v/v, 0.1% formic acid) for the HPLC analysis. Analytical standards of ($\pm$) abscisic acid (purity $\geq$ 98.5%) was purchased from Sigma-Aldrich; PA-grade methanol, acetone, dichloromethane and formic acid and HPLC-grade acetonitrile and water were purchased from VWR Chemicals (Milan, Italy). Analyses were performed with an Agilent 1260 Infinity HPLC (Santa Clara, CA, USA) equipped with a Sinergy 4 μm Hydro-RP (250 mm $\times$ 4.6 mm) column (Phenomenex, Castel Maggiore, Italy) with Security Guard at 35 °C, at a flow rate of 0.8 mL min^{-1}; the injection volume was 20 μL and the detection was made at 265 nm. The mobile phase of acetonitrile/water (30:70 v/v, 0.1% formic acid) was previously filtered and degassed. The compound was identified by comparing the retention times with those of the authentic reference compound. The peaks were quantified by an external standard method, using the measurements of the peak areas and a calibration curve. Stock solutions of ABA standards were prepared by diluting a solution (10 mg mL^{-1} in acetonitrile) to obtain a range of concentrations from 0.01 to 10 mg mL^{-1}. The limit of detection (LOD) was 0.005 mg L^{-1}.

4.5. Statistical Analysis

Light reflection and transmission were analysed by a one-way analysis of variance (ANOVA), using Sigma Stat 3.5 (Systat Software, San Jose, CA, USA). Treatment comparison was performed by Student's t-test at $p \leq 0.05$. Data taken over time were analysed with the repeated measure analysis of variance routine embedded in the XLSTAT 2019.1 software package (Addinsoft, Paris, France). Least squares mean method at $p < 0.05$ was used for multiple comparisons within dates. Correlation between parameters was tested by regression analyses and all models were calculated using Sigma Plot 11.0 (Systat Software, San Jose, CA, USA). R^2 significance was tested by ANOVA per $p = 0.05$.

5. Conclusions

Kaolin reduces the accumulation of ABA in the leaf by reducing the synthesis of Nx from Zx, resulting in faster recovery of vine gas exchange. These results further support the hypothesis that ABA mainly relates to its biosynthesis in leaves and that its accumulation can be limited by the downregulation of Nx synthesis. Moreover, this experiment provides the evidence that kaolin promotes the activity of the VAZ epoxidation/de-epoxidation cycle even under stressful conditions, thus preserving a full Vx pool after night recovery and a fluid energy dissipation of electron excess. Further experiments are needed to determine the biochemical pathway (gene expression and enzyme activity) leading to the downregulation of the Nx synthesis caused by kaolin.

Author Contributions: Conceptualization and planning, T.F. and S.T.; experimental work and analysis, T.F., C.S. and N.L.R.; writing—original draft preparation, S.T. and T.F.; planning and managing the plants used in the study, T.F.; writing—review and editing, S.P., P.S. and A.P.; data analysis, S.P. All authors have read and agreed to the published version of the manuscript.

Funding: This research received no external funding.

Acknowledgments: The authors wish to thank Massimo Benuzzi and Mauro Piergiacomi for topic discussions and Maria Giulia Parisi for technical support.

Conflicts of Interest: The authors declare no conflict of interest.

References

1. Adams, R.M.; Hurd, B.H.; Lenhart, S.; Leary, N. Effects of global climate change on agriculture: An interpretative review. *Clim. Res.* **1998**, *11*, 19–30. [CrossRef]
2. Olesen, J.E.; Bindi, M. Consequences of climate change for European agricultural productivity, land use and policy. *Eur. J. Agron.* **2002**, *16*, 239–262. [CrossRef]
3. Santillan, D.; Iglesias, A.; La Jeunesse, I.; Garrote, L.; Sotes, V. Vineyards in transition: A global assessment of the adaptation needs of grape producing regions under climate change. *Sci. Tot. Env.* **2009**, *657*, 830–852. [CrossRef]
4. Palliotti, A.; Tombesi, S.; Silvestroni, O.; Lanari, V.; Gatti, M.; Poni, S. Changes in vineyard establishment and canopy management urged by earlier climate-related grape ripening: A review. *Sci. Hortic.* **2004**, *178*, 43–54. [CrossRef]
5. Poni, S.; Lakso, A.N.; Turner, J.R.; Melious, R.E. Interactions of crop level and late season water stress on growth and physiology of field-grown Concord grapevines. *Am. J. Enol. Vitic.* **1994**, *45*, 252–258.
6. Flexas, J.; Bota, J.; Escalona, J.; Sampol, B.; Medrano, H. Effects of drought on photosynthesis in grapevines under field conditions: An evaluation of stomatal and mesophyll limitations. *Funct. Plant Biol.* **2002**, *29*, 461–471. [CrossRef]
7. Glenn, D.M.; Cooley, N.; Walker, R.; Clingeleffer, P.; Shellie, K. Impact of kaolin particle film and water deficit on wine grape water use efficiency and plant water relations. *HortScience* **2010**, *45*, 1178–1187. [CrossRef]
8. Glenn, D.M. The mechanisms of plant stress mitigation by kaolin-based particle films and applications in horticultural and agricultural crops. *HortScience* **2012**, *47*, 710–711. [CrossRef]
9. Brito, C.; Dinis, L.T.; Moutinho-Pereira, J.; Correia, C. Kaolin, an emerging tool to alleviate the effects of abiotic stresses on crop performance. *Sci. Hortic.* **2019**, *250*, 310–316. [CrossRef]
10. Rosati, A.; Metcalf, S.G.; Buchner, R.P.; Fulton, A.E.; Lampinen, B.D. Effects of kaolin application on light absorption and distribution, radiation use efficiency and photosynthesis of almond and walnut canopies. *Ann. Bot.* **2007**, *99*, 255–263. [CrossRef] [PubMed]
11. Cantore, V.; Pace, B.; Albrizio, R. Kaolin-based particle film technology affects tomato physiology, yield and quality. *Env. Exp. Bot.* **2009**, *66*, 279–288. [CrossRef]
12. Brillante, L.; Belfiore, N.; Gaiotti, F.; Lovat, L.; Sansone, L.; Poni, S.; Tomasi, D. Comparing Kaolin and pinolene to improve sustainable grapevine production during drought. *PLoS ONE* **2016**, *11*, e0156631. [CrossRef]
13. Dinis, L.T.; Ferreira, H.; Pinto, G.; Bernardo, S.; Correia, C.M.; Moutinho-Pereira, J. Kaolin-based, foliar reflective film protects photosystem II structure and function in grapevine leaves exposed to heat and high solar radiation. *Photosynthetica* **2016**, *54*, 47–55. [CrossRef]
14. Dinis, L.T.; Malheiro, A.C.; Luzio, A.; Fraga, H.; Ferreira, H.; Gonçalves, I.; Pinto, G.; Correia, C.M.; Moutinho-Pereira, J. Improvement of grapevine physiology and yield under summer stress by kaolin-foliar application: Water relations, photosynthesis and oxidative damage. *Photosynthetica* **2018**, *56*, 641–651. [CrossRef]
15. Frioni, T.; Saracino, S.; Squeri, C.; Tombesi, S.; Palliotti, A.; Sabbatini, P.; Magnanini, E.; Poni, S. Understanding kaolin effects on grapevine leaf and whole-canopy physiology during water stress and re-watering. *J. Plant Physiol.* **2019**, *242*, 153020. [CrossRef] [PubMed]

16. Brito, C.; Dinis, L.T.; Ferreira, H.; Rocha, L.; Pavia, I.; Moutinho-Pereira, J.; Correia, C.M. Kaolin particle film modulates morphological, physiological and biochemical olive tree responses to drought and rewatering. *Plant Phys. Biochem.* **2008**, *133*, 29–39. [CrossRef]

17. Nardini, A.; Salleo, S. Limitation of stomatal conductance by hydraulic traits: Sensing or preventing xylem cavitation? *Trees* **2000**, *15*, 14–24. [CrossRef]

18. Salleo, S.; Nardini, A.; Pitt, F.; Gullo, M.A. Xylem cavitation and hydraulic control of stomatal conductance in laurel (*Laurus nobilis* L.). *Plant Cell Environ.* **2000**, *23*, 71–79. [CrossRef]

19. Franks, P.J. Stomatal control and hydraulic conductance, with special reference to tall trees. *Tree Physiol.* **2004**, *24*, 865–878. [CrossRef]

20. Luan, S. Signalling drought in guard cells. *Plant Cell Environ.* **2002**, *25*, 229–237. [CrossRef]

21. Davies, W.J.; Kudoyarova, G.; Hartung, W. Long-distance ABA signaling and its relation to other signaling pathways in the detection of soil drying and the mediation of the plant's response to drought. *J. Plant Gro. Reg.* **2005**, *24*, 285–295. [CrossRef]

22. Brodribb, T.J.; McAdam, S.A.M. Passive origins of stomatal control in vascular plants. *Science* **2011**, *331*, 582–585. [CrossRef]

23. Xiong, L.; Zhu, J.K. Regulation of abscisic acid biosynthesis. *Plant Physiol.* **2003**, *133*, 29–36. [CrossRef]

24. Sack, L.; John, G.P.; Buckley, T.N. ABA accumulation in dehydrating leaves is associated with decline in cell volume, not turgor pressure. *Plant Physiol.* **2018**, *176*, 489–495. [CrossRef] [PubMed]

25. Arnoux, P.; Morosinotto, T.; Saga, G.; Bassi, R.; Pignol, D. A structural basis for the pH-dependent xanthophyll cycle in Arabidopsis thaliana. *Plant Cell* **2009**, *21*, 2036–2044. [CrossRef]

26. Jahns, P.; Miehe, B. Kinetic correlation of recovery from photoinhibition and zeaxanthin epoxidation. *Planta* **1996**, *198*, 202–210. [CrossRef]

27. Bouvier, F.; D'Harlingue, A.; Backhaus, R.A.; Kumagai, M.H.; Camara, B. Identification of neoxanthin synthase as a carotenoid cyclase paralog. *Eur. J. Biochem.* **2000**, *267*, 6346–6352. [CrossRef] [PubMed]

28. McAdam, S.A.M.; Sussmilch, F.C.; Brodribb, T.J. Stomatal responses to vapour pressure deficit are regulated by high speed gene expression in angiosperms. *Plant Cell Environ.* **2016**, *39*, 485–491. [CrossRef]

29. Dinis, L.T.; Bernardo, S.; Luzio, A.; Pinto, G.; Meijón, M.; Pintó-Marijuan, M.; Cotado, A.; Correira, C.; Moutinho-Pereira, J. Kaolin modulates ABA and IAA dynamics and physiology of grapevine under Mediterranean summer stress. *J. Plant Physiol.* **2018**, *220*, 181–192. [CrossRef]

30. Steiman, S.R.; Bittenbender, H.C.; Idol, T.W. Analysis of kaolin particle film use and its application on coffee. *HortScience* **2007**, *42*, 1605–1608. [CrossRef]

31. Wand, S.J.; Theron, K.I.; Ackerman, J.; Marais, S.J. Harvest and post-harvest apple fruit quality following applications of kaolin particle film in South African orchards. *Sci. Hortic.* **2006**, *107*, 271–276. [CrossRef]

32. Luciani, E.; Palliotti, A.; Frioni, T.; Tombesi, S.; Villa, F.; Zadra, C.; Farinelli, D. Kaolin treatments on Tonda Giffoni hazelnut (*Corylus avellana* L.) for the control of heat stress damages. *Sci. Hortic.* **2020**, *263*, 109097. [CrossRef]

33. Shellie, K.C.; King, B.A. Kaolin particle film and water deficit influence Malbec leaf and berry temperature, pigments, and photosynthesis. *Amer. J. Enol. Vitic.* **2013**, *64*, 223–230. [CrossRef]

34. Tombesi, S.; Nardini, A.; Frioni, T.; Soccolini, M.; Zadra, C.; Farinelli, D.; Poni, S.; Palliotti, A. Stomatal closure is induced by hydraulic signals and maintained by ABA in drought-stressed grapevine. *Sci. Rep.* **2015**, *5*, 12449. [CrossRef] [PubMed]

35. McAdam, S.A.M.; Brodribb, T.J. Mesophyll cells are the main site of abscisic acid biosynthesis in water-stressed leaves. *Plant Physiol.* **2018**, *177*, 911–917. [CrossRef] [PubMed]

36. Lorenz, D.H.; Eichhorn, K.W.; Bleiholder, H.; Klose, R.; Meier, U.; Weber, E. Growth stages of the grapevine: Phenological growth stages of the grapevine (*Vitis vinifera* L. ssp. *vinifera*)—Codes and descriptions according to the extended BBCH scale. *Austr. J. Grape Wine Res.* **1995**, *1*, 100–103. [CrossRef]

37. Poni, S.; Merli, M.C.; Magnanini, E.; Galbignani, M.; Bernizzoni, F.; Vercesi, A.; Gatti, M. An improved multichamber gas exchange system for determining whole-canopy water-use efficiency in grapevine. *Am. J. Enol. Vitic.* **2014**, *65*, 268–276. [CrossRef]

38. Gatti, M.; Pirez, F.J.; Frioni, T.; Squeri, C.; Poni, S. Calibrated, delayed-cane winter pruning controls yield and significantly postpones berry ripening parameters in *Vitis vinifera* L. cv. Pinot Noir. *Aust. J. Grape Wine Res.* **2018**, *24*, 305–316. [CrossRef]

39. Taylor, A.H. The measurement of diffuse reflection factors and a new absolute reflectometer. *J. Opt. Soc. Am.* **1920**, *4*, 9–23. [CrossRef]
40. Yuan, F.; Qian, M.C. Development of C13-norisoprenoids, carotenoids and other volatile compounds in *Vitis vinifera* L. Cv. Pinot noir grapes. *Food chem.* **2016**, *192*, 633–641. [CrossRef]
41. Vilaró, F.; Canela-Xandri, A.; Canela, R. Quantification of abscisic acid in grapevine leaf (*Vitis vinifera*) by isotope-dilution liquid chromatography–mass spectrometry. *Anal. Bioanal. Chem.* **2006**, *386*, 306–312. [CrossRef] [PubMed]

 International Journal of
Molecular Sciences

Article

The Role of ABA in Plant Immunity is Mediated through the PYR1 Receptor

Javier García-Andrade, Beatriz González, Miguel Gonzalez-Guzman, **Pedro L. Rodriguez** and **Pablo Vera ***

Instituto de Biología Molecular y Celular de Plantas, Universidad Politécnica de Valencia-C.S.I.C, Ciudad Politécnica de la Innovación, Edificio 8E, 46000 Valencia, Spain; jagarsercs@gmail.com (J.G.-A.); beagong1@ibmcp.upv.es (B.G.); mguzman@uji.es (M.G.-G.); prodriguez@ibmcp.upv.es (P.L.R.)
* Correspondence: vera@ibmcp.upv.es; Tel.: +34-963877884; Fax: +34-963877859

Received: 16 July 2020; Accepted: 9 August 2020; Published: 14 August 2020

Abstract: ABA is involved in plant responses to a broad range of pathogens and exhibits complex antagonistic and synergistic relationships with salicylic acid (SA) and ethylene (ET) signaling pathways, respectively. However, the specific receptor of ABA that triggers the positive and negative responses of ABA during immune responses remains unknown. Through a reverse genetic analysis, we identified that PYR1, a member of the family of PYR/PYL/RCAR ABA receptors, is transcriptionally upregulated and specifically perceives ABA during biotic stress, initiating downstream signaling mediated by ABA-activated SnRK2 protein kinases. This exerts a damping effect on SA-mediated signaling, required for resistance to biotrophic pathogens, and simultaneously a positive control over the resistance to necrotrophic pathogens controlled by ET. We demonstrated that PYR1-mediated signaling exerted control on a priori established hormonal cross-talk between SA and ET, thereby redirecting defense outputs. Defects in ABA/PYR1 signaling activated SA biosynthesis and sensitized plants for immune priming by poising SA-responsive genes for enhanced expression. As a trade-off effect, *pyr1*-mediated activation of the SA pathway blunted ET perception, which is pivotal for the activation of resistance towards fungal necrotrophs. The specific perception of ABA by PYR1 represented a regulatory node, modulating different outcomes in disease resistance.

Keywords: ABA; ethylene; pathogens; plant immunity; PYR1; salicylic acid

1. Introduction

Pathogen recognition triggers the altered accumulation of three major defense hormones: salicylic acid (SA), jasmonic acid (JA), and ethylene (ET). SA is essential for establishing resistance to many virulent biotrophic pathogens, especially as a component of systemic acquired resistance (SAR) [1,2], while JA and ET tend to be associated with resistance to fungal necrotrophic pathogens [3,4]. While JA and ET interact synergistically to activate certain disease responses, the JA and ET pathways act at least independently or even antagonistically with respect to the SA-dependent pathway [4,5]. Antagonistic interactions between SA and JA hormone signaling networks have been characterized [6–8]. JA levels decline soon after SA begins to accumulate [9]; this, therefore, suggests that, in response to a pathogen that can induce synthesis of both SA and JA, cross-talk is used by the plant to adjust the response in favor of the more effective pathway (i.e., the SA-mediated pathway). Similarly, SA acts antagonistically with ET [10–13], and their biosynthesis pathways can be mutually repressed [14,15]. More recently, Huang et al. [16] revealed a mechanism by which SA antagonizes ET signaling: the direct interaction of NPR1 (the core component of SA signaling) with EIN3 (the transcription factor mediating ET-responses) blocks transcription of EIN3-induced genes, and this interaction is further enhanced by SA. Therefore, tradeoffs between plant defenses against pathogens with different lifestyles must be strictly regulated [4,17], implying the fine-tuned deployment of conserved defense signals in different plant-pathogen interactions.

ABA is another major phytohormone involved in the regulation of a great variety of abiotic stress responses in plants. In addition, ABA assists in controlling many developmental and growth characteristics of plants, including seed germination and dormancy, leaf abscission, closure of stomata, or inhibition of fruit ripening [18]. ABA also controls the responses of plants to biotic stresses caused by a broad range of plant pathogens [19–22]. However, the ABA effect varies in different pathosystems, being the outcome influenced by the infection biology. ABA biosynthesis is required for effective disease resistance against necrotrophic fungal pathogens [23–25], whereas ABA has been shown to be involved in conferring susceptibility against bacterial diseases, with ABA-deficient mutants showing resistance enhancement [21,26,27]. In fact, some bacteria have acquired new virulence strategies for exploiting their host through the secretion of type III virulence effectors that promote enhancement of ABA levels in the infected plant [28–30]. Therefore, endogenous ABA synergizes with JA and exhibits a complex antagonistic relationship with SA during disease development [6,7,29]. Likewise, antagonistic interactions between components of the ABA and ET signaling pathways seem to modulate gene expression in response to biotic and abiotic stress (Fujimoto et al., 2000; Chen et al., 2002; Anderson et al., 2004; Yang et al., 2005; Broekaert et al., 2006) [5,31–34], but it remains unknown whether a convergent point exists between these two signaling pathways or whether they operate in parallel. Despite all these evidences, the specific components of the ABA signaling apparatus, which exploit the positive and negative responses of ABA during immune responses, remain unknown. Therefore, understanding the regulatory system of ABA-mediated responses to pathogens is critical for improving agricultural issues related to disease resistance. In contrast, specific components of ABA perception have been recently identified for stomatal closure signal integration [35]. Thus, PYL2 is sufficient for guard-cell ABA-induced response, and PYL4/5 are essential receptors for a guard-cell response to CO_2 [35].

Three major protein families form the core ABA signaling pathway; (i) the soluble ABA receptors, which are 14 members of pyrabactin resistance 1 (PYR1) and PYR1-like (PYL) proteins, also known as regulatory component of ABA receptors (RCAR) family and collectively referred to as PYR/PYL/RCAR, (ii) group A of type 2C protein phosphatases (PP2Cs), and (iii) SNF1-related protein kinases (SnRKs) subfamily 2 (SnRK2s), namely SnRK2.2, 2.3 and 2.6 (Cutler et al., 2010; Hubbard et al., 2010; Klingler et al., 2010; Raghavendra et al., 2010) [18,36–38]. In the absence of ABA, PP2Cs dephosphorylate and inactivate SnRK2s, repressing ABA-dependent responses [39,40]. When ABA concentration increases in response to stress conditions or developmental cues, ABA binds to receptors of the PYR/PYL/RCAR family, which leads to the formation of ternary complexes with PP2Cs, thereby inactivating them [41–43]. This results in the activation of SnRK2s, which subsequently phosphorylate a myriad of substrate proteins [44].

The PYR/PYL/RCAR ABA receptor family is unusually large, comprising 14 members in Arabidopsis and even more in crops, such as tomato, maize, or soybean (Gonzalez-Guzman et al., 2012; Gonzalez-Guzman et al., 2014; Helander et al., 2016) [45–47]. However, the biological roles of the individual PYR/PYL/RCAR members are still being established, which is complicated by functional redundancy. At least 13 PYR/PYL/RCAR members are able to perceive ABA, and the generation of quadruple, pentuple, and sextuple mutants is required to obtain robust ABA-insensitive phenotypes [41,43,45]. Moreover, the analysis of combined *pyr/pyl* mutants shows quantitative regulation of both stomatal aperture and transcriptional response to ABA [45]. Inactivation of six highly transcribed members, *PYR1*, *PYL1*, *PYL2*, *PYL4*, *PYL5*, and *PYL8*, generates a mutant that is practically blind to ABA in the classical assays that measure ABA sensitivity [45]. However, in spite of the receptor gene expression patterns and biochemical analyses of different receptor-phosphatase complexes suggesting that the function of ABA receptors is not completely redundant [45,48,49], only the single *pyl8* mutant has been reported to show a non-redundant role in root sensitivity to ABA [50]. In contrast, *pyr1* shows wild-type sensitivity to ABA and only shows a conditional phenotype -pyrabactin resistance in germination assays in medium supplemented with the ABA agonist pyrabactin [43]. In eukaryotes, functional diversification follows the evolutionary expansion of a gene family. Identification of specific roles for members of a multigene family is usually limited by laboratory conditions, whereas the plethora of conditions found in complex biological contexts offers chances to identify specific roles. Here, we were able to unveil a non-redundant role in plant immunity for PYR1, one of the 13 members of the multigene ABA receptor family, and revealed that the PYR1 receptor is pivotal in modulating the cross-talk between the SA and ET signaling pathways during the defense.

2. Results

2.1. The SnRK2s Protein Kinases are Engaged in Disease Resistance to Fungal Infection

Liquid chromatography-mass spectrometry (LC-MS) showed marked accumulation of ABA in full expanded leaves of Arabidopsis plants at 72 h after drop inoculation with a spore suspension of the fungal necrotroph *Plectosphaerella cucumerina* (Figure 1A). ABA enhancement supported the upregulation of *ABI4* gene expression, an ABA-responsive gene encoding a transcription factor [23] (Figure 1B). Therefore, ABA biosynthesis and signaling were triggered by *P. cucumerina* infection. The ABA-mediated activation of three monomeric SnRK2s (i.e., SnRK2.2, −2.3, and −2.6) is central to ABA signaling [51], so we investigated whether SnRK2s were engaged in the defense responses to this pathogen. Transgenic lines overexpressing HA-tagged SnRK2.6 (SnRK2.6-HA/OE) and SnRK2.2 (SnRK2.2-HA/OE) were inoculated with *P. cucumerina* or mock-treated, and leaf samples were collected at 0, 24, and 48 h post-inoculation (h.p.i.). Immunoprecipitation of SnRK2.2-HA and SnRK2.6-HA and the subsequent kinase assay of the immunoprecipitate were performed by determining the incorporation of ^{32}P to purified ABF2 protein fragment substrate (amino acids Gly-73 to Gln-119) [52] in gel-kinase assays. Results revealed two- and three-fold enhancement for SnRK2.6 and SnRK2.2 kinase activity, respectively, following fungal inoculation (Figure 1C,D). For both kinases, enhanced activity occurred at 24 h.p.i., and the activation was sustained at 48 h.p.i. Therefore, ABA-activated SnRK2s were actively engaged in response to this fungal pathogen.

We then investigated whether gain-of-function or loss-of-function in SnRK2s altered disease resistance to *P. cucumerina*. Symptoms of the fungal disease appear in the form of necrotic lesions, which are measured to quantify the degree of plant susceptibility [25,53,54]. Inoculation of transgenic plants individually overexpressing (OE) SnRK2.2, −2.3, and −2.6 revealed no significant variation in disease susceptibility towards *P. cucumerina* when compared to Col-0 plants (Figure 1E); thus, either endogenous SnRK2s levels are sufficient to achieve pathogen-triggered ABA signaling or overexpression of SnRK2s additionally requires increased ABA levels to enhance their activity. Although functional redundancy between SnRK2.2 and SnRK2.3 exists, functional segregation between SnRK2.6 and SnRK2.2/2.3 has been described [52]. Therefore, we inoculated an *snrk2.2/2.3* double mutant and the single *snrk2.6* mutant with *P. cucumerina* and recorded disease resistance. The triple *snrk2.2/2.3/2.6* mutant, which is drastically affected in plant growth [51], was not compatible with the pathogenic assay and was, therefore, not used in the present study. Figure 1F–G show that *snrk2.2/2.3* and *snrk2.6* plant resistance to *P. cucumerina* was severely compromised. Moreover, an ABA deficient mutant (i.e., *aba2*) was similarly affected in disease resistance to this pathogen (Figure 2A). In summary, our results indicated that pathogen-induced ABA accumulation and concurrent activation of SnRK2s positively regulated disease resistance to *P. cucumerina*.

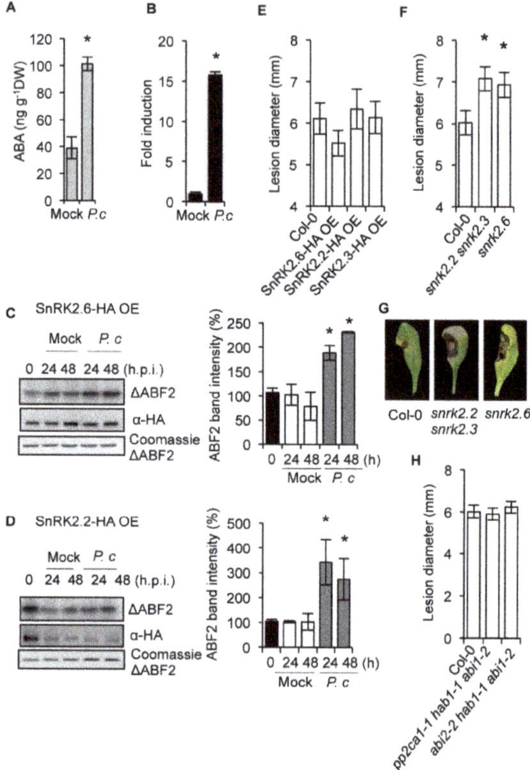

Figure 1. Participation of SnRK2s kinases in the response of Arabidopsis plants to infection by the fungal pathogen *P. cucumerina*. (**A**) ABA accumulation determined in mock and *P. cucumerina*-infected Col-0 plants. (**B**) RT-qPCR of *ABI4* in mock and in *P. cucumerina*-infected Col-0. (**C,D**) *P. cucumerina*-mediated activation of SnRK2.6 (**C**) and SnRK2.2 (**D**). Transgenic Arabidopsis plants expressing HA-tagged versions of the kinases were inoculated with *P. cucumerina*, or were mocked, and leaf samples were taken at 0, 24, and 48 h.p.i., and the protein extracts were immunoprecipitated with anti-HA antibodies. The immunoprecipitates were incubated with a His-ABF2 fragment (Gly73 to Gln 119; ΔABF2) in the presence of [γ-^{32}P]ATP, and the proteins were resolved by SDS-PAGE. Bands corresponding to ΔABF2 fragments and to SnRK2.6 and SnRK2.2 kinases are indicated. Radioactivities of ΔABF2 fragment bands were measured with a phosphoimager, and the values were plotted on the graphs shown at the right of the figures. Error bars indicate S.E.M.; n = 3. (**E**) Disease resistance towards *P. cucumerina* of transgenic plants overexpressing SnRK2.6, SnRK2.2, and SnRK2.3 in comparison to Col-0. (**F**) Disease resistance towards *P. cucumerina* in the double *snrk2.2 snrk2.3* mutant and in *snrk2.6* mutant plants. (**G**) Representative leaves from each genotype at 12 days following inoculation with *P. cucumerina*. (**H**) Disease resistance towards *P. cucumerina* in the triple PP2C mutants *pp2ca1 1hab1 1abi1-2* and *abi2-2 hab1 abi1-2*. For the bioassays with *P. cucumerina*, lesion diameter of 25 plants per genotype and four leaves per plant were determined 12 d following inoculation with *P. cucumerina*. Data points represent the average lesion size ± SE of measurements. An ANOVA was conducted to assess significant differences in the activation of SnRKs, ABA accumulation, ABI4 transcript accumulation, and disease symptoms, with a priori $p < 0.05$ level of significance; the asterisks * above the bars indicate statistically significant differences regarding mock treatments or Col-0 plants. Asterisks above the bars indicate different homogeneous groups with statistically significant differences.

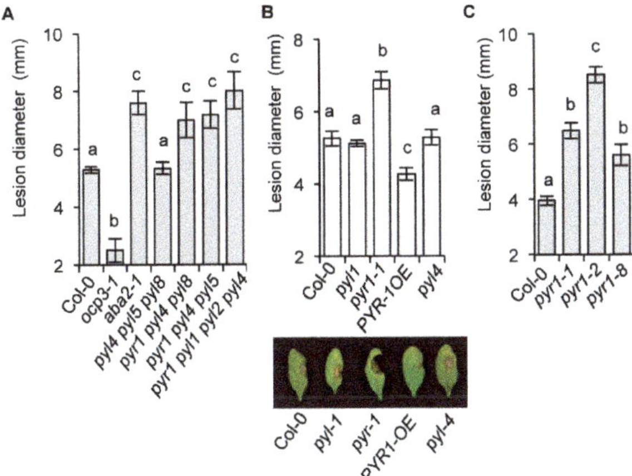

Figure 2. PYR1 is required for disease resistance towards *P. cucumerina*. (**A**). Disease resistance towards *P. cucumerina* in Col-0, the resistant *ocp3-1 mutant*, the susceptible *aba2-1*, and the triple and quadruple multi-locus mutants *pyl4 pyl5 pyl8*, *pyr1 pyl4 pyl8*, *pyr1 pyl4 pyl5*, and *pyr1 pyl1 pyl2 pyl4*. (**B**) Disease resistance in single *pyl1*, *pyr1*, and *pyl4* mutants, in a transgenic line overexpressing PYR1 (PYR1-OE), and in Col-0. Below the graph, the representative leaves from each genotype are shown at 12 days following inoculation with *P. cucumerina*. (**C**) Comparative disease resistance towards *P. cucumerina* among the allelic *pyr1-1*, *pyr1-2*, and *pyr1-8* mutants. Data points represent the average lesion size ± SE of measurements. An ANOVA was conducted to assess significant differences in disease symptoms ($p < 0.05$); the letters above the bars indicate different homogeneous groups with statistically significant differences.

ABA signaling through SnRK2s is negatively regulated by clade A protein phosphatase type 2C (PP2C), particularly by ABI1, ABI2, PP2CA/AHG3, AHG1, HAB1, and HAB2 (see [55] and references therein). Therefore, clade A PP2Cs might negatively regulate ABA-mediated disease resistance to *P. cucumerina*. Because of the demonstrated redundancy existing for these PP2Cs, combined inactivation of selected groups of these phosphatases is required to determine functionality. We combined loss-of-function mutations in ABI1, ABI2, HAB1, and PP2CA genes to determine their contribution to ABA-mediated disease resistance. Different combinations of mutations were used with two triple mutants, *pp2ca1-1;hab1-1;abi1-2* and *abi2-2;hab1-1;abi1-2*, which represent four of the nine closely related group A PP2Cs. Both multi-locus mutants showed an extreme response to exogenous ABA, partial constitutive response to endogenous ABA, and partial constitutive activation of SnRK2s in *pp2ca1-1;hab1-1;abi1-2* [51,55]. Inoculation of both triple mutants with *P. cucumerina* showed no defective disease resistance (Figure 1H). This result suggests that the demonstrated redundancy of PP2Cs masks the manifestation of a clear phenotype upon pathogen inoculation. Additionally, ABA response in triple *pp2c* mutants was partially equivalent to that of lines OE SnRK2s, which did not show altered disease resistance to the pathogen (Figure 1E). It is also possible that other members of the large PP2C family, represented by 76 homologous genes [56], are key for resistance to *P. cucumerina*. This interpretation is supported by previous studies showing that a distinct PP2C member (i.e., AtDBP1) is required for other aspects of plant immunity [57,58].

2.2. The Requirement of the PYR1 Receptor for Antifungal Resistance

We next investigated which one of 14 soluble PYR/PYL/RCAR receptors perceived the ABA produced during *P. cucumerina* infection. Partial functional redundancy of ABA receptors has been demonstrated by genetic analysis; however, PYL8 plays a non-redundant role to regulate root sensitivity to ABA [45,46]. Additionally, both transcriptional and physiological ABA responses and signaling of environmental cues in guard cells mediated by individual receptors are starting to be elucidated [35]. We characterized disease resistance to *P. cucumerina* in a series of multi-locus mutants from different PYR/PYL receptors. The triple *pyl4;pyl5;pyl8*, *pyr1;pyl4;pyl8*, and *pyr1;pyl4;pyl5* mutants, and the quadruple *pyr1;pyl1;pyl2;pyl4* mutant, representing the highest genetic impairment in PYR/PYL function without affecting plant growth [45], were inoculated with *P. cucumerina*, and their impact on disease resistance was compared to *aba2-1* (which enhances susceptibility [25]), to *ocp3-1* (which enhances resistance [53]), and Col-0 plants. The two triple mutants incorporating the *pyr1* mutation (i.e., *pyr1;pyl4;pyl8* and *pyr1;pyl4;pyl5*) exhibited noticeably enhanced disease susceptibility (Figure 2A), which was of a magnitude similar to that observed in *aba2-1* plants. Conversely, the disease resistance of the triple *pyl4;pyl5;pyl8* mutant was unaltered compared to Col-0 plants. The quadruple mutant (also containing the *pyr1* mutation) enhanced disease susceptibility to *P. cucumerina*. The results showed that the PYR1 receptor was pivotal for eliciting ABA-mediated defense responses towards *P. cucumerina*.

The specificity of PYR1 at eliciting plant immune responses was further tested by assaying the single *pyr1-1* mutant. The individual *pyr1-1* mutant had a compromised disease resistance phenotype (Figure 2B), contrasting to other single *pyl* mutants (e.g., *pyl1*, *pyl4*) for which resistance to the fungus remained intact. Moreover, the overexpression of the PYR1 receptor (PYR1-OE line) conferred significant enhancement of resistance to the fungus (Figure 2B). Other mutant alleles of the PYR1 receptor, predicted to produce a variety of defects in PYR1 (i.e., *pyr1-2* and *pyr1-8* [43]), consistently compromised disease resistance to *P. cucumerina*, showing *pyr1-2* mutant allele as the strongest phenotype (Figure 2C). These results supported that the PYR1 receptor positively promoted ABA-dependent plant immunity against *P. cucumerina*. Interestingly, these results also indicated that other major receptors for ABA response, i.e., PYL1, PYL4, PYL5, PYL8, were not recruited in plant response against *P. cucumerina*. Furthermore, PYR1 appeared similarly to be required for the immune activation to *Alternaria brassicicola*, another fungal necrotroph and the causal agent of black spot disease in Brassica species. Results shown in Supplemental Figure S1 indicate that upon inoculation with *A. brassicicola*, both *aba2* and *pyr1* plants, compared to Col-0, *pyl1*, and *pyl4* plants, showed remarkable enhancement in disease susceptibility to this pathogen. The enhancement of necrosis in *A. brassicicola*-inoculated leaves of *pyr1* plants gave further support to the importance of PYR1-mediated perception of ABA for mounting effective defense responses towards necrotrophs.

2.3. Local Induction of PYR1 Gene Expression by P. cucumerina

A reasonable explanation for the specific role of PYR1 in plant immunity might be the specific upregulation of *PYR1* expression in response to the pathogen. Therefore, we next investigated whether transcriptional reprogramming occurred to enhance *PYR1* expression upon pathogen inoculation. Transgenic plants expressing the promoter of the *PYR1* gene fused to the β-glucuronidase GUS reporter gene (*pPYR1::GUS*) [45] were used to detect potential *P. cucumerina*-mediated activation of *PYR1*. Transgenic lines carrying the *pPYL1::GUS* and *pPYL4::GUS* gene constructs were also assayed to determine specificity. Local infection, i.e., by drop inoculation on the upper leaf surface with a *P. cucumerina* spore suspension, of transgenic *pPYR1::GUS* plants revealed early transcriptional activation of *PYR1* triggered by the pathogen (Figure 3A). *PYR1* induction mostly occurred within the vascular bundles of the primary and secondary veins of the *P. cucumerina*-inoculated leaf sectors. This highly localized induced expression pattern was specific to *PYR1* because neither *PYL1* nor *PYL4* genes were transcriptionally activated under similar circumstances (Figure 3A). The local induction of *pPYR1::GUS* concurred with local synthesis and deposition of callose (Figure 3B) and later on with cell death (Figure 3C). These microscopy markers demarcated inoculated tissue sectors in advance to the appearance of visible necrosis and served to delimit local transcriptional responses. Moreover, callose deposition was compromised in *pyr1-1* and *aba2-1* mutants following fungal infection (Figure 3D), thus supporting the participation of ABA and PYR1 in this local process.

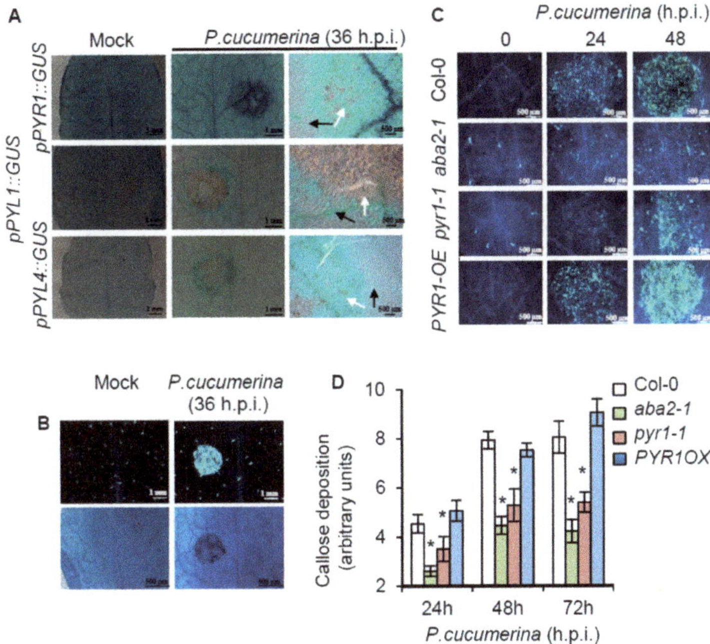

Figure 3. Local activation of *PYR1* gene expression at pathogen inoculation sites, and the requirement of PYR1 for pathogen-induced callose deposition. (**A**) Comparative histochemical analysis of GUS activity in rosette leaves from transgenic plants carrying *pPYR1::GUS*, *pPYL1::GUS*, and *pPYL4::GUS* gene constructs and those were either mocked or inoculated *P. cucumerina*. Leaves were stained for GUS activity at 36 h.p.i. The left panel corresponds to mocked plants. The central and right panels correspond to enlargements of the inoculated leaf sectors. Black arrow points towards leaf tissues proximal to the inoculation point, and white arrows denote tissues that directly received the spore inoculum. Note that *pPYR1::GUS* is heavily induced in leaf veins within the inoculated sector. (**B**) Characteristic spore-inoculated leaf sector, similar to those shown in A, stained with aniline blue to detect pathogen-induced callose deposition (top panel), or with trypan blue (lower panel) to identify incipient cell deterioration due to fungal infection at 36 h.p.i. (**C**) Aniline blue staining and epifluorescence microscopy were applied to visualize callose accumulation. Micrographs indicate *P. cucumerina* inoculation and infection site in the different Arabidopsis genotypes at 0 h.p.i (right panel), at 24 h.p.i. (central panel), and at 48 h.p.i. (right panel). (**D**) The number of yellows pixels (corresponding to pathogen-induced callose) per million on digital photographs of infected leaves were used as a means to express arbitrary units (i.e., to quantify the image) at the indicated times. Bars represent mean ± SD, *n* = 15 independent replicates. An ANOVA was conducted to assess significant differences in callose deposition ($p < 0.05$); the asterisks * above the bars indicate statistically significant differences regarding Col-0 plants.

2.4. Resistance Enhancement of pyr1 Plants to Pseudomonas syringae DC3000

In marked contrast to the results shown above, the role of ABA in repressing plant immunity against the (hemi) biotrophic pathogens *P. syringae* DC3000 has been previously documented [6,19–21]. Therefore, we asked whether the negative role of ABA in plant immunity against *P. syringae* DC3000 could similarly be funneled through PYR1. If so, we would expect resistance enhancement in *pyr1* plants. *pyr1* plants were inoculated by leaf infiltration with *P. syringae* DC3000, and the rate of bacterial growth in the inoculated leaves was determined at 3 days post-inoculation in comparison to *aba2*, *pyl1*, *pyl4*, and Col-0 plants. Figure 4A shows that bacterial growth was reduced 10-fold in both *aba2* and *pyr1* mutants compared to Col-0, *pyl1*, and *pyl4*. This result confirmed the negative role of ABA in resistance towards *P. syringae* DC3000 and demonstrated the specific requirement of PYR1 for the negative role of ABA during this plant-pathogen interaction. Moreover, pre-treatment of Col-0 with 150 μM ABA, applied by drenching, predictably provoked disease susceptibility enhancement to *P. syringae* DC3000 (Figure 4B), denoting a damping effect of ABA on SA signaling. This ABA-mediated enhancement in susceptibility to *P. syringae* DC3000 did not occur in *pyr1-2* plants whose enhanced resistance was not altered by the hormone (Figure 4B).

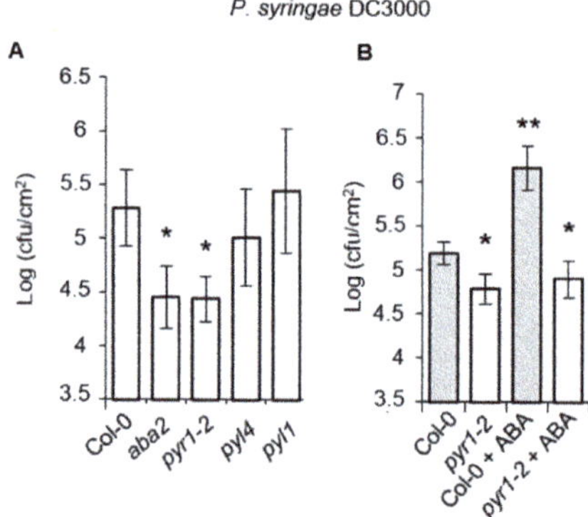

Figure 4. Response of *pyr1* plants to infection by *P. syringae* DC3000. (**A**) Col-0, *aba2-1*, *pyr1-2*, *pyl1*, and *pyl4* mutants were inoculated with *P. syringae* DC3000, and their disease responses were recorded. (**B**) Col-0 and *pyr1* plants were pre-treated with 150 μM ABA, applied by drenching, before inoculation with *P. syringae* DC3000, and the growth of the bacteria was recorded in comparison to mocked plants. Growth of *P. syringae* DC3000 was measured at 3 d.p.i. Error bars represent standard deviation (*n* = 12). An ANOVA was conducted to assess significant differences in disease symptoms, with a priori *p* < 0.05 level of significance; the asterisks *, ** above the bars indicate different homogeneous groups with statistically significant differences.

Therefore, our results indicated that the dual antagonistic role of ABA in plant immunity was mediated through the PYR1 receptor, which reciprocally activates and represses immune responses towards necrotrophic and biotrophic pathogens, respectively.

2.5. SA–Responsive Defense Genes are Activated in PYR1 Defective Mutants

We investigated whether *pyr1* and *aba2* plants carried constitutive elevated expression of SA-responsive genes, which might explain the observed enhanced resistance to *P. syringae* DC3000 (Figure 4). The accumulation of *PR-1* and *PR-2* transcript, which are SA- and pathogen-responsive genes, was examined by RT-qPCR. In addition, we examined *PR-4* and *PR-5*, which are also pathogen-responsive genes but are simultaneously influenced by SA and ET [59]. Transcript accumulation was also evaluated in *pyl1* and *pyl4* mutants, which served as additional controls. Figure 5A shows that *pyr1* and *aba2* plants carried constitutive elevated levels of SA-dependent *PR-1* and *PR-2* transcripts compared to Col-0, *pyl1*, or *pyl4* plants. Conversely, the constitutive levels of transcript accumulation for *PR-4* and *PR-5* occurring in Col-0 were repressed in *pyr1* and *aba2* plants, and only partially enhanced in *pyl1* plants (Figure 5A). The enhanced expression of *PR-1* and *PR-2* and the concerted repression of *PR-4* and *PR-5* were corroborated by the *pyr1* allelic series, with the *pyr1-2* allele showing the strongest differences (Figure 5B). Thus, the ABA/PYR1 module might function as an integration node regulating distinct branches of defenses. The constitutive activation of *PR1*- and -*2* in *pyr1* plants supported the enhanced accumulation of both free and conjugated SA observed in the mutant, which concurred also with elevated expression of *ICS1*, encoding isochorismate synthase, a pivotal enzyme controlling SA biosynthesis [60,61] (Figure 5C,D). On the other hand, *pyr1* plants only showed a moderate reduction, less than two-fold, of JA content in comparison to Col-0 (Figure 5E). The conspicuous enhancement in SA content in healthy *pyr1* plants, therefore, explained the resistance phenotypes of the mutant when confronted with *P. syringae* DC3000. However, the notorious enhanced susceptibility of *pyr1* plants to the fungal necrotrophs *P. cucumerina* and *A. brassicicola* remained unsolved, as it could not simply be explained by the moderate reduction of JA levels as attained in the mutant.

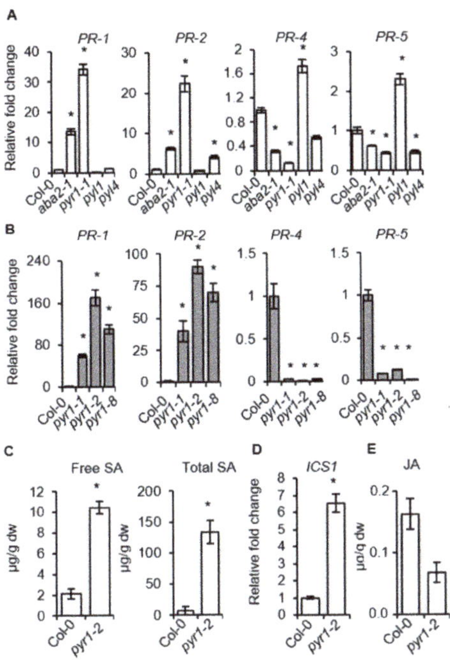

Figure 5. Expression of SA-responsive and ET-responsive genes in *pyr1* and *aba2* mutants. (**A**,**B**) RT-qPCR analysis showing constitutive expression levels of *PR-1*, *PR-2*, *PR-4*, and *PR-5* genes in (**A**) Col-0, *aba2-1*, *pyr1-1*, *pyl1*, and *pyl4* plants, and (**B**) their comparative expression levels in the allelic *pyr1-1*, *pyr1-2*, and *pyr1-8* mutants. Data represent mean ± SD; n = 3 replicates. The expression was normalized to the constitutive *ACT2* and *ACT8* genes and then to the expression in Col-0 plants. (**C–E**) Accumulation of free SA, total SA, and total JA in Col-0 and *pyr1-2* plants. Data represent the average of three biological replicates. An ANOVA was conducted to assess significant differences in RT-qPCR and hormone analysis, with a priori $p < 0.05$ level of significance; the asterisks * above the bars indicate statistically significant differences regarding Col-0 plants.

2.6. Enhanced Activation of MAPK Kinases in pyr1 Plants

We next investigated whether enhanced resistance to *P. syringae* DC3000 in *pyr1* plants was associated with elevated MAPKs activation, which is linked to the activation of immune responses following pathogen perception. We employed an antibody recognizing the phosphorylated residues within the MAPK activation loop (i.e., the pTEpY motif). Western blot analysis of protein extracts derived from healthy Col-0 and *pyr1* plants showed positive immunoreactive signals in two polypeptides corresponding to MPK6 and MPK3 (Beckers et al., 2009) (Figure 6), and the densitometric scanning of blots indicated that the MPK3 immunoreactive band was more intense in *pyr1* plants. Inoculation with *P. syringae* DC3000 promoted further activation-associated dual TEY phosphorylation of MPKs, which was noticeably higher for MPK3 in *pyr1* compared to Col-0 plants at 24 h.p.i. (Figure 6). At the latter stages of infection (i.e., 48 h.p.i.), the MPK activation was similar in Col-0 and *pyr1* plants.

Therefore, MPK activation may be prone to activation in plants defective of ABA perception through the PYR1 receptor. Indeed, partial pre-activation of MPK was reflected in detectable PR-1 protein accumulating in *pyr1* plants at time zero (Figure 6). This result was in agreement with the higher expression level of the *PR-1* gene determined by RT-qPCR (Figure 5A,B). Interestingly, inoculation with *P. syringae* DC3000 promoted the further accumulation of the PR-1 protein, which progressively increased over time to a much higher level in the *pyr1* mutant compared to Col-0 plants (Figure 6).

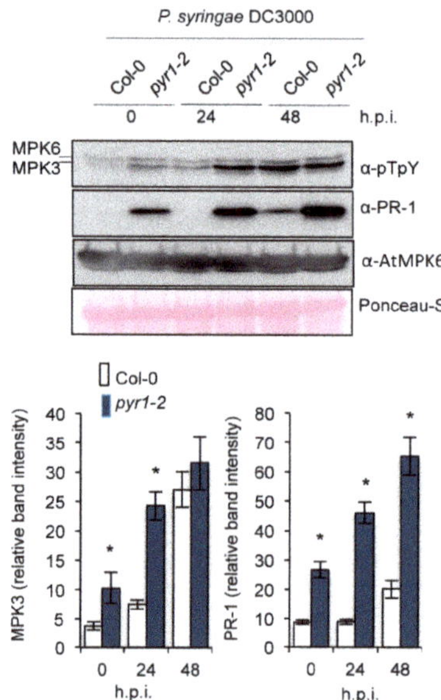

Figure 6. Loss of PYR1 function confers enhanced mitogen-activated kinase activation and PR-1 protein accumulation following *P. syringae* DC3000 infection. Western blot with anti-pTEpY and anti-PR-1 antibodies of crude protein extracts derived from Col-0, *pyr1-2* plants at 0, 24, and 48 h.p.i with *P. syringae* DC3000. Equal protein loading was check by Ponceau-S staining of the nitrocellulose filter. MPK6 and MPK3 migrating bands are indicated on the right. The experiments were repeated three times with similar results. Scan quantification of protein bands corresponding to MPK3 and PR-1 is shown below the Western blot. Data represent the mean ± SD; $n = 3$ replicates. An ANOVA was conducted to assess significant differences in RT-qPCR analysis, with a priori $p < 0.05$ level of significance; the asterisks * above the bars indicate statistically significant differences regarding Col-0 plants.

Thus, we hypothesize that the lack of ABA perception through the PYR1 receptor de-represses a pathway that allows cell sensitization through MPKs activation and downstream defense gene reprogramming, even in the absence of pathogen infection. Sensitized cells may be ready for the enhanced induction of this defense pathway following pathogen infection, which, in turn, may explain why *aba* and *pyr1* plants exhibit enhanced disease resistance to *P. syringae* DC3000. These observations support that ABA and PYR1 function as a repressor module of SA-mediated onset of resistance.

2.7. SA-Mediated Defense Genes are Poised for Enhanced Activation through Chromatin Remodeling in pyr1 Plants

We then asked whether other markers diagnostic of an immune status were also activated in *pyr1* plants. The expression of the extracellular subtilase *SBT3.3* gene has been recently described to be a switch for poising SA-related gene expression and immune priming [62]. Moreover, constitutive *SBT3.3* expression, MPK activation, and readied SA-related genes convey in plants defective in the RNA-directed DNA methylation (RdDM) pathway, which negatively regulates immune priming [54].

Consequently, the expression level of genes encoding SBT3.3 and either of the two subunits of RNA Pol V (i.e., NRPD2 and NRPE1) controlling RdDM were determined by RT-qPCR. Figure 7A shows the constitutive upregulation of *SBT3.3* and concurrent downregulation of *NRPD2* in *pyr1* plants compared to Col-0, congruent with the activation of immune priming in the mutant. The downregulation was specific for *NRPD2*, encoding the second large subunit of Pol V, because the expression of the gene encoding the large NRPE1 subunit exhibited a minimal variation in *pyr1* plants (Figure 7A).

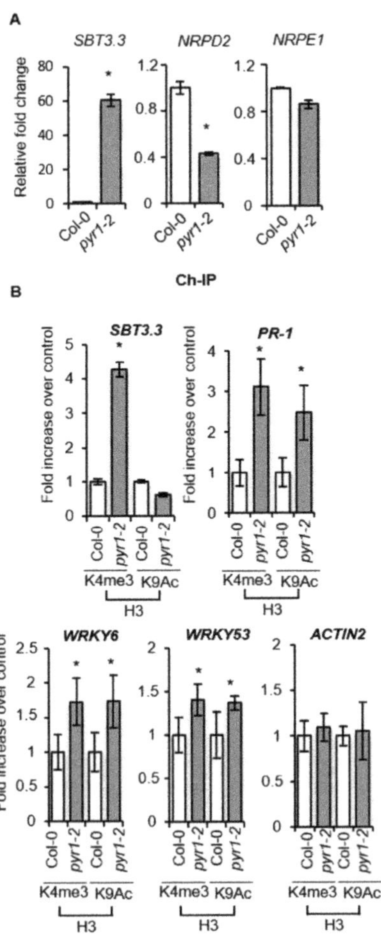

Figure 7. Loss of PYR1 function provokes the setting of hallmarks characteristic of primed immunity. (**A**) Comparative RT-qPCR of *SBT3.3*, *NRPD2*, and *NRPE1* transcript levels between healthy Col-0 and *pyr1-2* plants. The expression was normalized to the constitutive *ACT2/8* gene and then to the expression in Col-0 plants. (**B**) Chromatin immunoprecipitation (ChIP) and comparison between Col-0 and *pyr1-2* plants of the level of histone H3 Lys4 trimethylation (H3K4me3) and histone H3 Lys9 acetylation (H3K9ac) on the *SBT3.3*, *PR-1*, *WRKY6*, and *WRKY53* gene promoters as present in leaf samples. The setting of histone marks in *ACTIN2* was used as an internal control. Data are standardized for Col-0 histone modification levels. Data represent the mean ± SD; n = 3 biological replicates. An ANOVA was conducted to assess significant differences between MPKs activation and PR1 accumulation ($p < 0.05$); the asterisks * above the bars indicate statistically significant differences regarding Col-0 plants.

In plants defective in RdDM-mediated epigenetic control, immune priming is activated concurrently with chromatin histone activation marks being enriched in SA-related genes, including the *SBT3.3* gene itself [54,62]. Thus, we hypothesized that SA-related defense genes and *SBT3.3* in *pyr1-2* plants are poised for enhanced expression by differential histone modification. We used chromatin immunoprecipitation (ChIP) to analyze H3K4me3 and H3K9ac activation marks on the *SBT3.3* and *PR-1* gene promoter regions in *pyr1* and Col-0 plants. We also examined the genes encoding WRKY6 and WRKY53, transcriptional regulators of SA-defense genes. Figure 7B shows that H3K4me3 marks in the *SBT3.3* promoter region notably increased in *pyr1* plants compared to Col-0 plants, while H3K9ac marks remained invariant (Figure 7B), supporting previous descriptions of plants constitutively expressing primed immunity [54,62,63]. On the *PR1* promoter, both H3K4me3 and H3K9ac activation marks increased three- and two-fold, respectively, in *pyr1* plants compared to Col-0 (Figure 7B). Likewise, histone activation marks also moderately increased in the *WRKY6* and *WRKY53* promoters of *pyr1* plants compared to Col-0 plants (Figure 7B). The setting of histone marks in *pyr1* plants remained unchanged in the *ACTIN2* gene promoter, which was used as the control (Figure 7B). Therefore, chromatin activation marks proliferated in the promoter regions of the priming regulatory gene *SBT3.3* and the SA-responsive genes in *pyr1* plants and would explain why the PR-1 protein showed accelerated and enhanced accumulation in *pyr1* plants following pathogen inoculation (Figure 6). Our results indicated that ABA and its PYR1-mediated perception represented novel integral components of a signaling process, repressing SA-mediated immunity.

2.8. NahG Plants Abrogate the Altered Disease Resistance Response of pyr1 Plants

To evaluate the role of SA for *pyr1*-altered resistance, we generated a *pyr1;NahG* double mutant. In plants carrying the *NahG* transgene, salicylate hydroxylase depletes the plant of this defense hormone [64]. Compared to Col-0, *NahG* plants showed an anticipated increase in susceptibility to *P. syringae* DC3000 due to SA depletion (Figure 8A). Interestingly, in *pyr1;NahG* plants, the *pyr1*-mediated enhanced resistance was abrogated, and instead enhanced susceptibility to *P. syringae* DC3000 emerged (Figure 8A). Moreover, when assayed against the fungal pathogen *P. cucumerina, NahG* plants behaved like Col-0, both showing the same degree of susceptibility (Figure 8B), suggesting normal metabolic levels of SA played no major role in the resistance towards this pathogen. Surprisingly, in *pyr1;NahG* plants, the *pyr1*-mediated-enhanced susceptibility to *P. cucumerina* was abrogated (Figure 8B), with *pyr1;NahG* plants to be behaving as Col-0 or *NahG* plants. This suggested that the PYR1-mediated perception of ABA negatively regulated the SA pathway. When this negative regulation failed, such as in *pyr1* plants, the SA levels increased, and the resistance to *P. syringae* DC3000 was activated. As a trade-off effect, the elevated SA levels presumably interfered with JA or ET signaling pathways required for mounting a resistance response to fungal pathogens.

2.9. pyr1-Mediated Enhanced SA Content Blocks ET Perception

The SA and JA signal pathways are under an antagonistic equilibrium. Therefore, we wondered if the enhanced SA levels of *pyr1* plants could be affecting JA signaling in this mutant. We studied *pyr1* plants for altered responses to JA using the widely applied root growth inhibition assay. In the absence of JA, primary root length of *pyr1* seedlings was comparable to that of Col-0 plants (Supplemental Figure S2), and in the presence of JA, root growth reduction in the mutant was also similar to that observed in Col-0 plants (Supplemental Figure S2), providing evidence that JA perception was not impaired in the mutant. In addition, comparison of the expression level of different JA-responsive genes at different times following *P. cucumerina* inoculation in Col-0 and *pyr1* plants revealed that JA signaling appeared to be not affected in the mutant (Supplemental Figure S3). Instead, for some of the genes analyzed, a higher induction was recorded in *pyr1* plants. Therefore, JA signaling was not compromised in *pyr1* plants.

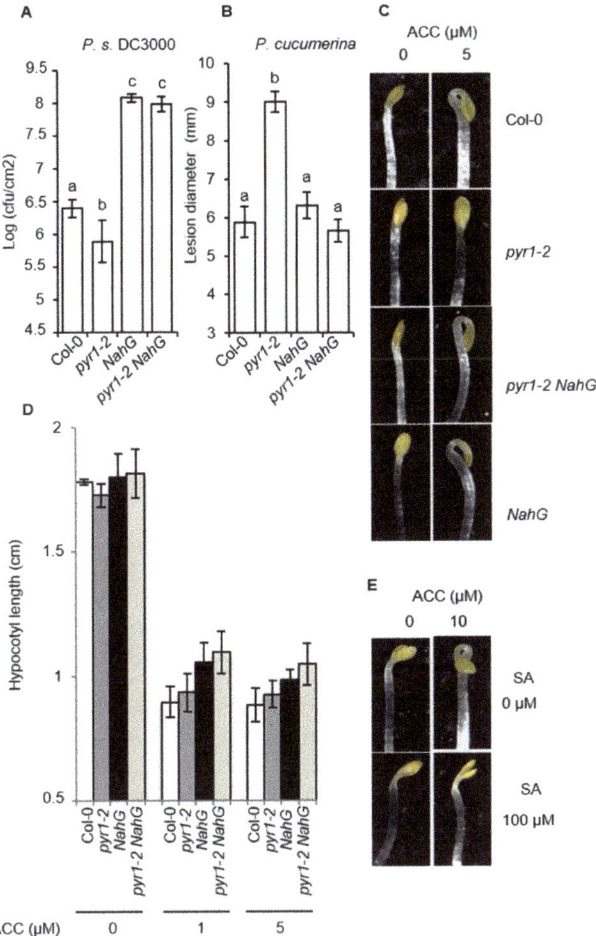

Figure 8. Effect of *NahG* on disease resistance and insensitivity to ACC of *pyr1* plants and seedlings. (**A,B**) Comparative disease resistance towards *P.s.* DC3000 and *P. cucumerina* among Col-0, *pyr1-2*, *NahG*, and *pyr1-2NahG* plants. Growth of *P. syringae* DC3000 was measured at 3 d.p.i. Error bars represent standard deviation ($n = 12$). For *P. cucumerina*, data points represent the average lesion size ± SE of measurements. An ANOVA was conducted to assess significant differences in disease symptoms ($p < 0.05$); the letters above the bars indicate different homogeneous groups with statistically significant differences. (**C**) Apical hook region of the indicated seedlings germinated and grown on MS/2 in the dark for 4 d in the presence of the indicated concentration of ACC. (**D**) Hypocotyl length of seedlings germinated and grown in the dark for 4 d on MS/2 medium supplemented with the denoted concentrations of ACC. Error bars represent standard deviation ($n = 50$). An ANOVA was conducted, and no significant differences were observed in hypocotyl length ($p < 0.05$). (**E**) Apical hook region of Col-0 seedlings germinated and grown on MS/2 in the dark for 4 d in the presence of the indicated concentration of ACC and SA.

We next asked whether ET signaling, which is also pivotal for resistance to fungal pathogens [10,12,13], could be the one impaired in *pyr1* plants due to the elevated levels of SA. This hypothesis gained even more relevance in view of the recently described mechanism explaining the antagonism between SA and ET in the suppression of apical hook formation and early seedling establishment via NPR1-mediated repression of EIN3 and EIL1 [16]. We, therefore, assayed Col-0 and *pyr1* seedlings, grown in the dark in the presence or absence of a low concentration of the ethylene precursor ACC (5 µM), for the induction of the ET-mediated triple response. The triple response in Arabidopsis consists of shortening and thickening of hypocotyls and roots and exaggeration of the curvature of apical hooks. Compared to Col-0, the assay revealed that *pyr1* seedlings showed no curvature of the apical hook (Figure 8C) and also showed less shortened hypocotyls (Figure 8D) when grown in the presence of ACC. Thus, the *pyr1* mutant was impaired in ET perception. The enhanced SA content in *pyr1* seedlings was the causal link mediating insensitivity to ET since in *pyr1;NahG* double mutant, normal sensitivity to ET was re-established (Figure 8C,D). Moreover, when Col-0 seedlings were assayed in the presence of high amounts of SA (100 µM), the ACC-induced triple response was abrogated (Figure 8E), further sustaining that the elevated levels of SA in *pyr1* plants blunted ET perception. Thus, our results suggested that perception of ABA through PYR1 acted primarily as a module negatively controlling the SA pathway. When ABA/PYR1 failed, the SA pathway was released, and the resistance to *P. syringae* DC3000 was activated. As a trade-off effect, the enhanced accumulation of the SA pathway blocked the ET pathway, the later required for resistance to fungal necrotrophic pathogens.

3. Discussion

Despite the demonstrated role of ABA on the final outcome of immune responses, the specific components of the ABA signaling apparatus and the specific mechanisms that exploit ABA to positively and negatively influence immune responses to specific plant-pathogen interactions have remained largely unknown. Here, we showed that SnRK2s kinases were actively engaged in activating resistance towards *P. cucumerina*, whereas the loss-of-function of any of the three individual SnRK2s compromised this resistance. Furthermore, we demonstrated that PYR1 was pivotal and played a positive role in disease resistance to *P. cucumerina* since overexpression of PYR1 (i.e., *PYR1-OE* transgenic line) conferred significantly enhanced resistance. Conversely, in PYR1 loss-of-function mutants, the resistance was compromised. Therefore, the PYR1 receptor had functional specificity in perceiving ABA produced in response to fungal infection to activate plant immunity. This study provided novel information about a specific ABA receptor-mediating specific plant immune responses and pinpointed ABA-activated SnRK2s as cardinal components for plant resistance. This information helps construct a functional classification scheme of the different members of the PYR/PYL receptor family with respect to their downstream signaling pathways in a true biological context. Thus, specific non-redundant roles for PYR1 and PYL8 have been reported in plant immunity (this work) and root ABA sensitivity [50], respectively. An explanation for the specific role of PYR1 in pathogen response could be the selective and highly localized pathogen-induced expression of *PYR1* in vascular bundles (Figure 3A). This expression pattern mirrors the expression of genes encoding ABA-biosynthetic enzymes [65–68]. Therefore, the synthesis of ABA and the pathogen-induced expression of *PYR1* spatially concur in the vasculature, supporting the hypothesis that vascular tissues function as an integrating node, triggering stress signaling that sets in motion the local and systemic immune responses in the plant [67,69,70].

This study showed that resistance to *A. brassicicola* was also dependent on ABA and PYR1, reinforcing the importance of this signal pathway for activating immunity against necrotrophs. This further reconciled with results shown above and also with previous studies showing that ABA promotes enhanced resistance to the necrotroph *P. cucumerina* [23,25,71]. Moreover, when a fungal necrotroph is a shift to a biotrophic lifestyle by changing the inoculation method and also the developmental stage of the plant [72], as reported for *P. cucumerina* [22], then ABA exerts an opposite effect, and the resistance to this same pathogen is suppressed. This contradictory role of ABA at controlling the disease resistance has also been observed for biotrophic pathogens (e.g., *P. syringae* DC3000), with resistance appearing negatively regulated by ABA, whereas resistance is enhanced in ABA-deficient mutants (Mohr & Cahill, 2007; Jensen et al., 2008; Fan et al., 2009; Verhage et al., 2010) [4,21,26,27]. In fact, we showed that the growth of *P. syringae* DC3000 was severely restricted in *pyr1* plants, as documented for *aba2 aao3* plants or the ABA-insensitive *abi1-1* and *abi2-1* mutants [8,28].

These results demonstrated the Janus functions of PYR1 in disease resistance, mediating repression of immunity against biotrophic pathogens, whereas activation against necrotrophs. Consequently, PYR1 may regulate which of these two plant immune programs prevails. This hypothesis supports previous observations of ABA as a hormone that interacts antagonistically or synergistically with the SA-JA-ET backbone of the plant immune signaling network, redirecting defense outputs [4,28,73–75]. Yet, how does the ABA/PYR1 module interfere with immunity to drive simultaneously the repression and activation of the SA and JA/ET defense pathways, respectively? Hormone cross-talk allows different hormone signaling pathways to act antagonistically or synergistically, providing the powerful regulatory potential to flexibly tailor the plant's adaptive response to a range of environmental cues [4]. Our results showed basal activation of the SA-dependent pathway in *pyr1* mutants, and that *pyr1* was insensitive to the damping effect of ABA on SA signaling. This finding supported previous work demonstrating the negative role of ABA on disease resistance to biotrophs, and that *P. syringae*-induced ABA levels in Col-0 suppress SA biosynthesis and action, enhancing susceptibility to this pathogen [21,28,73,75,76]. Interestingly, our finding that JA perception remained intact in *pyr1* plant but ET perception became compromised added a degree of specificity for the understanding of the disease resistance phenotype of the mutant. The observation that in *pyr1;NahG* plants, the *pyr1*-mediated ET-insensitivity was reversed, and that SA *per se* could block ET perception in Col-0 plants (Figure 8 and Huang et al., 2020), pointed towards SA-mediated repression of ET signaling modulated by ABA and PYR1 during pathogenesis. The positive effect of ABA at promoting ET-dependent resistance to fungal pathogens may be indirect: perception of pathogenic ABA by PYR1 dampens SA signaling, which, in turn, stops ET pathway repression by SA. This ABA and PYR1-modulated cross-talk regulation of SA and ET pathways may provide the plant with a powerful regulatory potential to boost its defenses according to the lifestyle of the attacker. This phenomenon may also explain why disease-promoting biotrophic pathogens (e.g., *P. syringae* DC3000) have developed strategies to alter the host ABA physiology as part of the infection strategy [28,76].

How does then ABA/PYR1-mediated signaling control SA-mediated defenses? *pyr1* plants bear constitutive activation of *ICS* expression and moderate enhanced level of SA. Besides, *pyr1* plants carry the hallmarks of immune priming, including (1) basal activation of MPKs; (2) repression of *NRPD2* and, therefore, the RdDM mechanisms that negatively control the onset of defense; (3) activation of the SBT3.3 subtilase; and (4) readying of SA-related genes for enhanced expression by pertinent chromatin modifications. Therefore, *pyr1* plants mirror the phenotypes of RdDM defective mutants, which exhibit simultaneous enhanced susceptibility and resistance to necrotrophs and biotrophs, respectively [54], supporting SA signaling activation in *pyr1* plants. The fact that immune priming and SA-mediated resistance are negatively regulated by the RdDM, and that in *pyr1* plants, the ABA repression of SA pathway is relieved, both observations unveil the importance of ABA/PYR1 as new element participating in an epigenetic mechanism of control of gene expression in plant immunity.

4. Materials and Methods

4.1. Plants Growth Conditions

Arabidopsis thaliana plants were grown in a growth chamber (19–23 °C, 85% relative humidity, $100 \text{ mEm}^{-2} \text{ s}^{-1}$ fluorescent illumination) on a 10-h-light and 14-hr-dark cycle. All mutants and transgenic plants are in Col-0 background; SnRK2.6-HA/OE and SnRK2.2-HA/OE were previously described [77,78]; snrk2.2 snrk2.3 and snrk2.6 were described in [44,45,51]; the triple mutants pp2ca1-1 hab1-1 abi2-2 and abi2-2 hab1-1 abi1-2 were described in [22]; ocp3-1 and aba2-1 mutants described in [25,53], and the triple pyl4 pyl5 pyl8, pyr1 pyl4 pyl8, and pyr1 pyl4 pyl5 mutants, along with the quadruple pyr1 pyl1 pyl2 pyl4 and single pyl1, pyl4, pyr1-1, pyr1-2 and pyr1-8 mutants were described in [41,45]. Transgenic lines carrying pPYR1::GUS, pPYL1::GUS, pPYL4::GUS were described previously [45]. NahG plants were described previously [64].

4.2. Gene Expression Analysis

Total RNA was extracted from plant tissues using TRIzol (Invitrogen, Waltham, MA, USA) and purified by lithium chloride precipitation. Reverse transcription was done using the RevertAid H Minus First Strand cDNA Synthesis Kit (Fermentas Life Sciences, Waltham, MA, USA). Quantitative PCR (qPCR) was performed using an ABI PRISM 7000 sequence detection system and SYBR-Green (Perkin-Elmer Applied Biosystems, Foster, CA, USA). *ACTIN2* and *ACTIN8* were the reference genes. The primers used for RT-qPCR experiments are provided in Table S1. RT-qPCR analyses were performed at least three times using sets of cDNA samples from independent experiments.

4.3. Immunoprecipitation of HA–SnRKs and In Vitro Phosphorylation

HA-tagged SnRK2.2 and SnRK2.6 were immunoprecipitated and used for in vitro kinase assay, as described previously [78].

4.4. Chromatin Immunoprecipitation

Chromatin isolation and immunoprecipitation were performed, as described [54,62]. Chip samples, derived from three biological replicates, were amplified in triplicate and measured by quantitative PCR using primers for *SBT3.3*, *PR-1*, *WRKY6*, *WRKY53*, and *Actin2*, as reported [54,62]. All ChIP experiments were performed in three independent biological replicates. The antibodies used for the immunoprecipitation of modified histones from 2 g of leaf material were antiH3K4m3 (#07-473 Millipore) and antiH3K9ac (#07-352 Millipore).

4.5. Western Blot

Protein crude extracts were prepared by homogenizing ground frozen leaf material with Tris-buffered saline (TBS) supplemented with 5 mM DTT, protease inhibitor cocktail (Sigma-Aldrich), and protein phosphatase inhibitors (PhosStop, Roche). Protein concentration was measured using Bradford reagent; 25 µg of total protein was separated by SDS-PAGE (12% acrylamide *w/v*) and transferred to nitrocellulose filters. The filter was stained with Ponceau-S after transfer and used as a loading control.

4.6. Pathogen Assays

Pseudomonas syringae DC3000 was grown for two days, and a culture with O.D. 2×10^{-4} was used to infect 5-week-old *Arabidopsis* leaves by infiltration, and the bacterial growth was determined following [54,62]. Twelve samples were used for each data point and represented as the mean ± SD of log c.f.u./cm^2. For *Plectosphaerella cucumerina* and *Alternaria brassicicola* bioassays, 5-week-old plants were inoculated, as described [23,24], with a suspension of fungal spores of 2.5×10^4, 5×10^6, and 5×10^6 spores/mL, respectively. The challenged plants were maintained at 100% relative humidity. Disease symptoms were evaluated by determining the lesion diameter of at least 100 lesions (25 plants per genotype and four leaves per plant) at 3, 12, and 8 days after inoculation with *P. cucumerina* and *A. brassicicola*, respectively. For pathogen-induced callose deposition analyses, infected leaves were stained at 24, 48, and 72 h.p.i. with aniline blue, and callose deposition quantifications were performed, as described by [53].

4.7. Determination of Plant Hormones and Metabolites

ABA, JA, SA levels were determined, as described previously [25,71].

Supplementary Materials: The following are available online at http://www.mdpi.com/1422-0067/21/16/5852/s1, Figure S1: Disease resistance towards *Alternaria brassicicola* in Col-0, *aba2-1*, *pyr1*, *pyl2* and *pyl4* plants and comparison of Col-0 and *pyr1-2* plants upon inoculation with *Altenaria brassicicola*. Figure S2: Response of Col-0 and *pyr1-2* seedlings to JA. Figure S3: Comparative RT-qPCR analysis for the expression of different JA-responsive and biosynthesis genes in either mocked or *P. cucumerina*-inoculated Col-0 and *pyr1* plants. Table S1: primer sequences.

Author Contributions: M.G.-G. and P.L.R. performed SnRKs kinase assays and provided genetic resources. J.G.-A. and B.G. performed the rest of the experiment shown in the manuscript. P.V. wrote the manuscript. All authors have read and agreed to the published version of the manuscript.

Funding: This research was founded by the Spanish AEI agency by grant BIO2017-82503-R to P.L.R. and by grant RTI2018-098501-B-I00 to P.V.

Acknowledgments: We acknowledge V. Flors for the technical assistance in the analysis of plant metabolites and hormones.

Conflicts of Interest: Authors declare no competing financial interest.

References

1. Malamy, J.; Carr, J.P.; Klessig, D.F.; Raskin, I. Salicylic acid: A likely endogenous signal in the resistance response of tobacco to viral infection. *Science* **1990**, *250*, 1002–1004. [CrossRef] [PubMed]
2. Métraux, J.P.; Signer, H.; Ryals, J.; Ward, E.; WyssBenz, M.; Gaudin, J.; Raschdorf, K.; Schmid, E.; Blum, W.; Inverardi, B. Increase in salicylic acid at the onset of systemic acquired resistance in cucumber. *Science* **1990**, *250*, 1004–1006. [CrossRef] [PubMed]
3. Glazebrook, J. Contrasting mechanisms of defense against biotrophic and necrotrophic pathogens. *Annu. Rev. Phytopathol.* **2005**, *43*, 205–227. [CrossRef]
4. Verhage, A.; van Wees, S.C.M.; Pieterse, C.M.J. Plant immunity: It's the hormones talking, but what do they say? *Plant Physiol.* **2010**, *154*, 536–540. [CrossRef] [PubMed]
5. Broekaert, W.F.; Delaure, S.L.; De Bolle, M.F.; Cammue, B.P. The role of ethylene in host-pathogen interactions. *Annu. Rev. Phytopathol.* **2006**, *44*, 393–416. [CrossRef]
6. Grant, M.R.; Jones, J.D. Hormone (dis)harmony moulds plant health and disease. *Science* **2009**, *324*, 750–752. [CrossRef]
7. Pieterse, C.M.J.; Van der Does, D.; Zamioudis, C.; Leon-Reyes, A.; Van Wees, S.C. Hormonal modulation of plant immunity. *Annu. Rev. Cell Dev. Biol.* **2012**, *28*, 489–521. [CrossRef]
8. Thaler, J.S.; Bostock, R.M. Interactions between abscisic-acid-mediated responses and plant resistance to pathogens and insects. *Ecology* **2004**, *85*, 48–58. [CrossRef]
9. Spoel, S.H.; Koornneef, A.; Claessens, S.M.; Korzelius, J.P.; Van Pelt, J.A.; Mueller, M.J.; Buchala, A.J.; Metraux, J.P.; Brown, R.; Kazan, K.; et al. NPR1 modulates cross-talk between salicylate- and jasmonate-dependent defense pathways through a novel function in the cytosol. *Plant Cell* **2003**, *15*, 760–770. [CrossRef]
10. Thomma, B.P.H.J.; Eggermont, K.; Tierens, K.F.M.; Broekaert, W.F. Requirement of functional *ethylene-insensitive 2* gene for efficient resistance of Arabidopsis to infection by *Botrytis cinerea*. *Plant Physiol.* **1999**, *121*, 1093–1102. [CrossRef]
11. Gu, Y.Q.; Yang, C.; Thara, V.K.; Zhou, J.; Martin, G.B. *Pti4* is induced by ethylene and salicylic acid, and its product is phosphorylated by the Pto kinase. *Plant Cell* **2000**, *12*, 771–786. [CrossRef] [PubMed]
12. Berrocal-Lobo, M.; Molina, A.; Solano, R. Constitutive expression of *ETHYLENE-RESPONSE-FACTOR1* in *Arabidopsis* confers resistance to several necrotrophic fungi. *Plant J.* **2002**, *29*, 23–32. [CrossRef] [PubMed]
13. Díaz, J.; ten Have, A.; van Kan, J.A.L. The role of ethylene and wound signaling in resistance of tomato to *Botrytis cinerea*. *Plant Physiol.* **2002**, *129*, 1341–1351. [CrossRef] [PubMed]
14. Leslie, C.A.; Romani, R.J. Inhibition of ethylene biosynthesis by salicylic acid. *Plant Physiol.* **1988**, *88*, 833–837. [CrossRef] [PubMed]
15. Chen, H.; Xue, L.; Chintamanani, S.; Germain, H.; Lin, H.; Cui, H.; Cai, R.; Zuo, J.; Tang, X.; Li, X.; et al. ETHYLENE INSENSITIVE3 and ETHYLENE INSENSITIVE3-LIKE1 repress *SALICYLIC ACID INDUCTION DEFICIENT2* expression to negatively regulate plant innate immunity in *Arabidopsis*. *Plant Cell* **2009**, *21*, 2527–2540. [CrossRef]
16. Huang, P.; Dong, Z.; Guo, P.; Zhang, X.; Qiu, Y.; Li, B.; Wang, Y.; Guo, H. Salicylic Acid Suppresses Apical Hook Formation via NPR1-Mediated Repression of EIN3 and EIL1 in Arabidopsis. *Plant Cell* **2020**, *32*, 612–629. [CrossRef] [PubMed]
17. Spoel, S.H.; Johnson, J.S.; Dong, X. Regulation of tradeoffs between plant defenses against pathogens with different lifestyles. *Proc. Natl. Acad. Sci. USA* **2007**, *104*, 18842–18847. [CrossRef]

18. Cutler, S.R.; Rodriguez, P.L.; Finkelstein, R.R.; Abrams, S.R. Abscisic acid: Emergence of a core signaling network. *Annu. Rev. Plant Biol.* **2010**, *61*, 651–679. [CrossRef]

19. Mauch-Mani, B.; Mauch, F. The role of abscisic acid in plant-pathogen interactions. *Curr. Opin. Plant Biol.* **2005**, *8*, 409–414. [CrossRef]

20. Asselbergh, B.; De Vleesschauwer, D.; Hofte, M. Global switches and fine-tuning-aba modulates plant pathogen defense. *Mol. Plant Microbe Interact.* **2008**, *21*, 709–719. [CrossRef]

21. Fan, J.; Hill, L.; Crooks, C.; Doerner, P.; Lamb, C. Abscisic acid has a key role in modulating diverse plant-pathogen interactions. *Plant Physiol.* **2009**, *150*, 1750–1761. [CrossRef] [PubMed]

22. Sanchez-Vallet, A.; Lopez, G.; Ramos, B.; Delgado-Cerezo, M.; Riviere, M.P.; Llorente, F.; Fernandez, P.V.; Miedes, E.; Estevez, J.M.; Grant, M.; et al. Disruption of abscisic acid signaling constitutively activates arabidopsis resistance to the necrotrophic fungus plectosphaerella cucumerina. *Plant Physiol.* **2012**, *160*, 2109–2124. [CrossRef] [PubMed]

23. Ton, J.; Mauch-Mani, B. Beta-amino-butyric acid-induced resistance against necrotrophic pathogens is based on aba-dependent priming for callose. *Plant J.* **2004**, *38*, 119–130. [CrossRef] [PubMed]

24. Adie, B.A.; Perez-Perez, J.; Perez-Perez, M.M.; Godoy, M.; Sanchez-Serrano, J.J.; Schmelz, E.A.; Solano, R. ABA is an essential signal for plant resistance to pathogens affecting JA biosynthesis and the activation of defenses in Arabidopsis. *Plant Cell* **2007**, *19*, 1665–1681. [CrossRef]

25. García-Andrade, J.; Ramírez, V.; Flors, V.; Vera, P. Arabidopsis ocp3 mutant reveals a mechanism linking aba and ja to pathogen-induced callose deposition. *Plant J.* **2011**, *67*, 783–794.

26. Mohr, P.G.; Cahill, D.M. Suppression by ABA of salicylic acid and lignin accumulation and the expression of multiple genes, in arabidopsis infected with *Pseudomonas syringae* pv. tomato. *Funct. Integr. Genomics* **2007**, *7*, 181–191. [CrossRef]

27. Jensen, M.K.; Hagedorn, P.H.; de Torres-Zabala, M.; Grant, M.R.; Rung, J.H.; Collinge, D.B.; Lyngkjaer, M.F. Transcriptional regulation by an nac (nam-ataf1,2-cuc2) transcription factor attenuates aba signalling for efficient basal defence towards Blumeria graminis f. sp. Hordei in Arabidopsis. *Plant J.* **2008**, *56*, 867–880. [CrossRef]

28. Torres-Zabala, M.; Truman, W.; Bennett, M.H.; Lafforgue, G.; Mansfield, J.W.; Rodriguez Egea, P.; Bögre, L.; Grant, M. *Pseudomonas syringae* pv. tomato hijacks the *Arabidopsis* abscisic acid signalling pathway to cause disease. *EMBO J.* **2007**, *26*, 1434–1443. [CrossRef]

29. Mine, A.; Berens, M.L.; Nobori, T.; Anver, S.; Fukumoto, K.; Winkelmüller, T.M.; Takeda, A.; Becker, D.; Tsuda, K. Pathogen exploitation of an abscisic acid- and jasmonate-inducible MAPK phosphatase and its interception by *Arabidopsis* immunity. *Proc. Natl. Acad. Sci. USA* **2017**, *114*, 7456–7461. [CrossRef]

30. Peng, Z.; Hu, Y.; Zhang, J.; Huguet-Tapia, J.; Block, A.K.; Park, S.; Sapkota, S.; Liu, Z.; Liu, S.; White, F.F. *Xanthomonas translucens* commandeers the host rate-limiting step in ABA biosynthesis for disease susceptibility. *Proc. Natl Acad. Sci. USA* **2019**, *116*, 20938–20946. [CrossRef]

31. Fujimoto, S.Y.; Ohta, M.; Usui, A.; Shinshi, H.; Ohme-Takagi, M. Arabidopsis ethylene-responsive element binding factors act as transcriptional activators or repressors of gcc box-mediated gene expression. *Plant Cell* **2000**, *12*, 393–404. [PubMed]

32. Chen, W.; Provart, N.J.; Glazebrook, J.; Katagiri, F.; Chang, H.S.; Eulgem, T.; Mauch, F.; Luan, S.; Zou, G.; Whitham, S.A.; et al. Expression profile matrix of arabidopsis transcription factor genes suggests their putative functions in response to environmental stresses. *Plant Cell* **2002**, *14*, 559–574. [CrossRef] [PubMed]

33. Anderson, J.P.; Badruzsaufari, E.; Schenk, P.M.; Manners, J.M.; Desmond, O.J.; Ehlert, C.; Maclean, D.J.; Ebert, P.R.; Kazan, K. Antagonistic interaction between abscisic acid and jasmonate-ethylene signaling pathways modulates defense gene expression and disease resistance in arabidopsis. *Plant Cell* **2004**, *16*, 3460–3479. [CrossRef] [PubMed]

34. Yang, Z.; Tian, L.; Latoszek-Green, M.; Brown, D.; Wu, K. Arabidopsis erf4 is a transcriptional repressor capable of modulating ethylene and abscisic acid responses. *Plant Mol. Biol.* **2005**, *58*, 585–596. [CrossRef]

35. Dittrich, M.; Mueller, H.M.; Bauer, H.; Peirats-Llobet, M.; Rodriguez, P.L.; Geilfus, C.M.; Carpentier, S.C.; Al Rasheid, K.A.S.; Kollist, H.; Merilo, E.; et al. The role of Arabidopsis ABA receptors from the PYR/PYL/RCAR family in stomatal acclimation and closure signal integration. *Nat. Plants* **2019**, *5*, 1002–1011. [CrossRef]

36. Hubbard, K.E.; Nishimura, N.; Hitomi, K.; Getzoff, E.D.; Schroeder, J.I. Early abscisic acid signal transduction mechanisms: Newly discovered components and newly emerging questions. *Genes Dev.* **2010**, *24*, 1695–1708. [CrossRef]

37. Klingler, J.P.; Batelli, G.; Zhu, J.K. Aba receptors: The start of a new paradigm in phytohormone signalling. *J. Exp. Bot.* **2010**, *61*, 3199–3210. [CrossRef]

38. Raghavendra, A.S.; Gonugunta, V.K.; Christmann, A.; Grill, E. ABA perception and signalling. *Trends Plant Sci.* **2010**, *15*, 395–401. [CrossRef]

39. Umezawa, T.; Sugiyama, N.; Mizoguchi, M.; Hayashi, S.; Myouga, F.; Yamaguchi-Shinozaki, K.; Ishihama, Y.; Hirayama, T.; Shinozaki, K. Type 2c protein phosphatases directly regulate abscisic acid-activated protein kinases in arabidopsis. *Proc. Natl. Acad. Sci. USA* **2009**, *106*, 17588–17593. [CrossRef]

40. Vlad, F.; Rubio, S.; Rodrigues, A.; Sirichandra, C.; Belin, C.; Robert, N.; Leung, J.; Rodriguez, P.L.; Lauriere, C.; Merlot, S. Protein phosphatases 2c regulate the activation of the snf1-related kinase ost1 by abscisic acid in arabidopsis. *Plant Cell* **2009**, *21*, 3170–3184. [CrossRef]

41. Fujii, H.; Chinnusamy, V.; Rodrigues, A.; Rubio, S.; Antoni, R.; Park, S.-Y.; Cutler, S.R.; Sheen, J.; Rodriguez, P.L.; Zhu, J.-K. In vitro reconstitution of an aba signaling pathway. *Nature* **2009**, *462*, 660–664. [CrossRef] [PubMed]

42. Ma, Y.; Szostkiewicz, I.; Korte, A.; Moes, D.; Yang, Y.; Christmann, A.; Grill, E. Regulators of pp2c phosphatase activity function as abscisic acid sensors. *Science* **2009**, *324*, 1064–1068. [CrossRef] [PubMed]

43. Park, S.Y.; Fung, P.; Nishimura, N.; Jensen, D.R.; Fujii, H.; Zhao, Y.; Lumba, S.; Santiago, J.; Rodrigues, A.; Chow, T.F.; et al. Abscisic acid inhibits type 2c protein phosphatases via the pyr/pyl family of start proteins. *Science* **2009**, *324*, 1068–1071. [CrossRef] [PubMed]

44. Umezawa, T.; Sugiyama, N.; Takahashi, F.; Anderson, J.C.; Ishihama, Y.; Peck, S.C.; Shinozaki, K. Genetics and phosphoproteomics reveal a protein phosphorylation network in the abscisic acid signaling pathway in arabidopsis thaliana. *Sci. Signal* **2013**, *6*, rs8. [CrossRef]

45. Gonzalez-Guzman, M.; Pizzio, G.A.; Antoni, R.; Vera-Sirera, F.; Merilo, E.; Bassel, G.W.; Fernandez, M.A.; Holdsworth, M.J.; Perez-Amador, M.A.; Kollist, H.; et al. Arabidopsis pyr/pyl/rcar receptors play a major role in quantitative regulation of stomatal aperture and transcriptional response to abscisic acid. *Plant Cell* **2012**, *24*, 2483–2496. [CrossRef]

46. Gonzalez-Guzman, M.; Rodriguez, L.; Lorenzo-Orts, L.; Pons, C.; Sarrion-Perdigones, A.; Fernandez, M.A.; Peirats-Llobet, M.; Forment, J.; Moreno-Alvero, M.; Cutler, S.R.; et al. Tomato pyr/pyl/rcar abscisic acid receptors show high expression in root, differential sensitivity to the abscisic acid agonist quinabactin, and the capability to enhance plant drought resistance. *J. Exp. Bot.* **2014**, *65*, 4451–4464. [CrossRef]

47. Helander, J.D.; Vaidya, A.S.; Cutler, S.R. Chemical manipulation of plant water use. *Bioorg. Med. Chem.* **2016**, *24*, 493–500. [CrossRef]

48. Antoni, R.; Gonzalez-Guzman, M.; Rodriguez, L.; Rodrigues, A.; Pizzio, G.A.; Rodriguez, P.L. Selective inhibition of clade a phosphatases type 2c by pyr/pyl/rcar abscisic acid receptors. *Plant Physiol.* **2012**, *158*, 970–980. [CrossRef]

49. Tischer, S.V.; Wunschel, C.; Papacek, M.; Kleigrewe, K.; Hofmann, T.; Christmann, A.; Grill, E. Combinatorial interaction network of abscisic acid receptors and coreceptors from arabidopsis thaliana. *Proc. Natl. Acad. Sci. USA* **2017**, *114*, 10280–10285. [CrossRef]

50. Antoni, R.; Gonzalez-Guzman, M.; Rodriguez, L.; Peirats-Llobet, M.; Pizzio, G.A.; Fernandez, M.A.; De Winne, N.; De Jaeger, G.; Dietrich, D.; Bennett, M.J.; et al. Pyrabactin resistance1-like8 plays an important role for the regulation of abscisic acid signaling in root. *Plant Physiol.* **2013**, *161*, 931–941. [CrossRef]

51. Fujii, H.; Zhu, J.K. Arabidopsis mutant deficient in 3 abscisic acid-activated protein kinases reveals critical roles in growth, reproduction, and stress. *Proc. Natl. Acad. Sci. USA* **2009**, *106*, 8380–8385. [CrossRef] [PubMed]

52. Fujii, H.; Verslues, P.E.; Zhu, J.K. Identification of two protein kinases required for abscisic acid regulation of seed germination, root growth, and gene expression in arabidopsis. *Plant Cell* **2007**, *19*, 485–494. [CrossRef] [PubMed]

53. García-Andrade, J.; Ramirez, V.; Lopez, A.; Vera, P. Mediated plastid rna editing in plant immunity. *PLoS Pathog.* **2013**, *9*, e1003713. [CrossRef] [PubMed]

54. López, A.; Ramírez, V.; García-Andrade, J.; Flors, V.; Vera, P. The RNA silencing enzyme RNA polymerase V is required for plant immunity. *PLoS Genet.* **2011**, *7*, e1002434. [CrossRef] [PubMed]

55. Rubio, S.; Rodrigues, A.; Saez, A.; Dizon, M.B.; Galle, A.; Kim, T.H.; Santiago, J.; Flexas, J.; Schroeder, J.I.; Rodriguez, P.L. Triple loss of function of protein phosphatases type 2c leads to partial constitutive response to endogenous abscisic acid. *Plant Physiol.* **2009**, *150*, 1345–1355. [CrossRef] [PubMed]

56. Schweighofer, A.; Hirt, H.; Meskiene, I. Plant PP2C phosphatases: Emerging functions in stress signaling. *Trends Plant Sci.* **2004**, *9*, 236–243. [CrossRef]

57. Carrasco, J.L.; Ancillo, G.; Mayda, E.; Vera, P. A novel transcription factor involved in plant defense endowed with protein phosphatase activity. *EMBO J.* **2003**, *22*, 3376–3384. [CrossRef]

58. Carrasco, J.L.; Castello, M.J.; Naumann, K.; Lassowskat, I.; Navarrete-Gomez, M.; Scheel, D.; Vera, P. Arabidopsis protein phosphatase dbp1 nucleates a protein network with a role in regulating plant defense. *PLoS ONE* **2014**, *9*, e90734. [CrossRef]

59. Clarke, J.D.; Volko, S.M.; Ledford, H.; Ausubel, F.M.; Dong, X. Roles of salicylic acid, jasmonic acid, and ethylene in cpr-induced resistance in Arabidopsis. *Plant Cell* **2000**, *12*, 2175–2190. [CrossRef]

60. Wildermuth, M.C.; Dewdney, J.; Wu, G.; Ausubel, F.M. Isochorismate synthase is required to synthesize salicylic acid for plant defence. *Nature* **2001**, *414*, 562–565. [CrossRef]

61. Garcion, C.; Lohmann, A.; Lamodière, E.; Catinot, J.; Buchala, A.; Doermann, P.; Métraux, J.-P. Characterization and biological function of the isochorismate synthase2 gene of Arabidopsis. *Plant Physiol.* **2008**, *147*, 1279–1287. [CrossRef] [PubMed]

62. Ramírez, V.; López, A.; Mauch-Mani, B.; Gil, M.J.; Vera, P. An extracellular subtilase switch for immune priming in arabidopsis. *PLoS Pathog.* **2013**, *9*, e1003445. [CrossRef]

63. Mauch-Mani, B.; Baccelli, I.; Luna, E.; Flors, V. Defense priming: An adaptive part of induced resistance. *Annu. Rev. Plant Biol.* **2017**, *68*, 485–512. [CrossRef]

64. Gaffney, T.; Friedrich, L.; Vernooij, B.; Negrotto, D.; Nye, G.; Uknes, S.; Ward, E.; Kessmann, H.; Ryals, J. Requirement of salicylic acid for the induction of systemic acquired resistance. *Science* **1993**, *261*, 754–756. [CrossRef] [PubMed]

65. Cheng, W.H.; Endo, A.; Zhou, L.; Penney, J.; Chen, H.C.; Arroyo, A.; Leon, P.; Nambara, E.; Asami, T.; Seo, M.; et al. A unique short-chain dehydrogenase/reductase in arabidopsis glucose signaling and abscisic acid biosynthesis and functions. *Plant Cell* **2002**, *14*, 2723–2743. [CrossRef] [PubMed]

66. Koiwai, H.; Nakaminami, K.; Seo, M.; Mitsuhashi, W.; Toyomasu, T.; Koshiba, T. Tissue-specific localization of an abscisic acid biosynthetic enzyme, AAO3, in Arabidopsis. *Plant Physiol.* **2004**, *134*, 1697–1707. [CrossRef]

67. Endo, A.; Koshiba, T.; Kamiya, Y.; Nambara, E. Vascular system is a node of systemic stress responses: Competence of the cell to synthesize abscisic acid and its responsiveness to external cues. *Plant Signal Behav.* **2008**, *3*, 1138–1140. [CrossRef]

68. Endo, A.; Sawada, Y.; Takahashi, H.; Okamoto, M.; Ikegami, K.; Koiwai, H.; Seo, M.; Toyomasu, T.; Mitsuhashi, W.; Shinozaki, K.; et al. Drought induction of arabidopsis 9-cis-epoxycarotenoid dioxygenase occurs in vascular parenchyma cells. *Plant Physiol.* **2008**, *147*, 1984–1993. [CrossRef]

69. Alvarez, M.E.; Pennell, R.I.; Meijer, P.J.; Ishikawa, A.; Dixon, R.A.; Lamb, C. Reactive oxygen intermediates mediate a systemic signal network in the establishment of plant immunity. *Cell* **1998**, *92*, 773–784. [CrossRef]

70. Yu, A.; Lepere, G.; Jay, F.; Wang, J.; Bapaume, L.; Wang, Y.; Abraham, A.L.; Penterman, J.; Fischer, R.L.; Voinnet, O.; et al. Dynamics and biological relevance of DNA demethylation in Arabidopsis antibacterial defense. *Proc. Natl. Acad. Sci. USA* **2013**, *110*, 2389–2394. [CrossRef]

71. Flors, V.; Ton, J.; van Doorn, R.; Jakab, G.; Garcia-Agustin, P.; Mauch-Mani, B. Interplay between JA, SA and ABA signalling during basal and induced resistance against pseudomonas syringae and alternaria brassicicola. *Plant J.* **2008**, *54*, 81–92. [CrossRef] [PubMed]

72. Petriacq, P.; Stassen, J.H.; Ton, J. Spore density determines infection strategy by the plant pathogenic fungus plectosphaerella cucumerina. *Plant Physiol.* **2016**, *170*, 2325–2339. [CrossRef] [PubMed]

73. Koornneef, A.; Pieterse, C.M. Cross talk in defense signaling. *Plant Physiol.* **2008**, *146*, 839–844. [CrossRef]

74. Ton, J.; Flors, V.; Mauch-Mani, B. The multifaceted role of aba in disease resistance. *Trends Plant Sci.* **2009**, *14*, 310–317. [CrossRef]

75. De Vleesschauwer, D.; Yang, Y.; Cruz, C.V.; Hofte, M. Abscisic acid-induced resistance against the brown spot pathogen cochliobolus miyabeanus in rice involves map kinase-mediated repression of ethylene signaling. *Plant Physiol.* **2010**, *152*, 2036–2052. [CrossRef] [PubMed]

76. de Torres Zabala, M.; Bennett, M.H.; Truman, W.H.; Grant, M.R. Antagonism between salicylic and abscisic acid reflects early host-pathogen conflict and moulds plant defence responses. *Plant J.* **2009**, *59*, 375–386. [CrossRef]

77. Belin, C.; de Franco, P.O.; Bourbousse, C.; Chaignepain, S.; Schmitter, J.M.; Vavasseur, A.; Giraudat, J.; Barbier-Brygoo, H.; Thomine, S. Identification of features regulating ost1 kinase activity and ost1 function in guard cells. *Plant Physiol.* **2006**, *141*, 1316–1327. [CrossRef]
78. Planes, M.D.; Ninoles, R.; Rubio, L.; Bissoli, G.; Bueso, E.; Garcia-Sanchez, M.J.; Alejandro, S.; Gonzalez-Guzman, M.; Hedrich, R.; Rodriguez, P.L.; et al. A mechanism of growth inhibition by abscisic acid in germinating seeds of arabidopsis thaliana based on inhibition of plasma membrane H+-atpase and decreased cytosolic pH, K+, and anions. *J. Exp. Bot.* **2015**, *66*, 813–825. [CrossRef]

International Journal of
Molecular Sciences

Article

The Arabidopsis RLCK VI_A2 Kinase Controls Seedling and Plant Growth in Parallel with Gibberellin

Ildikó Valkai [1], Erzsébet Kénesi [1], Ildikó Domonkos [1], Ferhan Ayaydin [1], Danuše Tarkowská [2], Miroslav Strnad [2], Anikó Faragó [3,4], László Bodai [3] and Attila Fehér [1,5,*]

1   Institute of Plant Biology, Biological Research Centre, 6726 Szeged, Hungary; ildiko.valkai@gmail.com (I.V.); kenesi.erzsebet@brc.hu (E.K.); domonkos.ildiko@brc.hu (I.D.); ayaydin.ferhan@brc.hu (F.A.)
2   Laboratory of Growth Regulators, Institute of Experimental Botany, Czech Academy of Sciences and Palacky University, 78371 Olomouc, Czech Republic; tarkowska@ueb.cas.cz (D.T.); miroslav.strnad@upol.cz (M.S.)
3   Department of Biochemistry and Molecular Biology, Faculty of Science and Informatics, University of Szeged, 6726 Szeged, Hungary; f.ancsi93@freemail.hu (A.F.); bodai@bio.u-szeged.hu (L.B.)
4   Doctoral School in Biology, Faculty of Science and Informatics, University of Szeged, 6726 Szeged, Hungary
5   Department of Plant Biology, Faculty of Science and Informatics, University of Szeged, 6726 Szeged, Hungary
*   Correspondence: feher.attila@brc.hu

Received: 3 September 2020; Accepted: 29 September 2020; Published: 1 October 2020

Abstract: The plant-specific receptor-like cytoplasmic kinases (RLCKs) form a large, poorly characterized family. Members of the RLCK VI_A class of dicots have a unique characteristic: their activity is regulated by Rho-of-plants (ROP) GTPases. The biological function of one of these kinases was investigated using a T-DNA insertion mutant and RNA interference. Loss of RLCK VI_A2 function resulted in restricted cell expansion and seedling growth. Although these phenotypes could be rescued by exogenous gibberellin, the mutant did not exhibit lower levels of active gibberellins nor decreased gibberellin sensitivity. Transcriptome analysis confirmed that gibberellin is not the direct target of the kinase; its absence rather affected the metabolism and signalling of other hormones such as auxin. It is hypothesized that gibberellins and the RLCK VI_A2 kinase act in parallel to regulate cell expansion and plant growth. Gene expression studies also indicated that the kinase might have an overlapping role with the transcription factor circuit (PIF4-BZR1-ARF6) controlling skotomorphogenesis-related hypocotyl/cotyledon elongation. Furthermore, the transcriptomic changes revealed that the loss of *RLCK VI_A2* function alters cellular processes that are associated with cell membranes, take place at the cell periphery or in the apoplast, and are related to cellular transport and/or cell wall reorganisation.

Keywords: *Arabidopsis thaliana*; cell expansion; gibberellins; hypocotyl growth; transcriptomic analysis; plant hormones; plant size; receptor-like cytoplasmic kinase; skotomorphogenesis

1. Introduction

Eukaryotic protein kinases form a large superfamily, and are associated essentially with all cellular functions. During evolution, protein kinase families have evolved independently in the lineages of eukaryotes. In consequence, the various kinase families are unevenly represented in the different eukaryotic organisms, which also have specific kinase classes. The number of protein kinase coding genes is especially high in plant genomes. In *Arabidopsis thaliana*, ~4% of the protein coding genes code for protein kinases while this percentage is ca. 2% in *Homo sapiens* [1]. This high number of protein kinases is likely due to the importance of cell-to-cell communication during the post-embryonic development of plants that is strongly influenced by the environment. In addition, plant defence and immunity depend on the specific recognition of pathogen-associated molecular patterns. In plants, cell-to-cell communication as well as innate immunity rely on plant receptor-like kinases (RLKs), which account for more than half of protein kinases in Arabidopsis (>600 RLK genes out of >1000 kinase-coding genes). RLKs resemble the receptor kinases of animals. The sequence of their kinase domain indicates that they are rather related to the animal cytoplasmic Pelle and interleukin receptor-associated kinases [2]. Moreover, RLKs exhibit serine/threonine kinase specificity in contrast to animal receptor kinases that are almost exclusively tyrosine kinases. This indicates that ancient RLK/Pelle kinases were co-opted for transmembrane signalling in plants after their divergence from animals.

RLKs can be classified into several families based on their various extracellular ligand-binding and slightly divergent cytoplasmic kinase domains [2]. There are also a number of RLK-like kinases that have only cytoplasmic kinase domain but no extracellular ligand-binding domain (only a few of them have a transmembrane domain) [3]. These cytoplasmic protein kinases are referred to as receptor-like cytoplasmic kinases, or receptor-like cytoplasmic kinases (RLCKs). The 149 RLCKs of Arabidopsis were divided into 17 subfamilies (RLCK-II and RLCK-IV to RLCK-XIX), based on sequence homology [3]. RLCKs often associate with RLKs to mediate cellular signalling in response to various RLK-sensed environmental and/or developmental signals [4]. Most of the RLCKs, however, have unknown functions.

The Arabidopsis RLCK-VI family is divided into two groups, with seven members each: RLCK VI_A and RLCK VI_B [5]. RLCK VI_A but not RLCK VI_B kinases were shown to bind plant Rho-type small GTPases (ROPs) in their GTP-bound state [6–9]. This binding results in augmented in vitro kinase activity [6,8,9]. Regulation of kinase activity by Rho-type GTPases is well known in animal and yeast cells as well. In these organisms, the kinase classes regulated by Rho-type Rho/Rac/Cdc42 GTPases are the p21-activated kinases (PAKs), the Rho-kinases (ROKs), the mixed-lineage kinases (MLKs), the myotonin-related Cdc42-binding kinases (MRCKs), the citron kinases (CRIKs), and the novel protein kinase (PKN) [10]. However, plant genomes code for none of these kinases [11]. It seems that, during the evolution of land plants, a sub-group of plant-specific RLCKs were co-opted to mediate ROP GTPase signalling [12]. While the GTPase-binding ability of yeast and animal Rho-type GTPase-regulated kinases is due to the presence of defined structural elements outside of their kinase domains (such as the Cdc42/Rac-interactive binding -CRIB- motif of PAKs), RLCK VI_A kinases use conserved amino acids widely distributed in the kinase domain to form a binding surface for ROPs [12].

At present, the members of the RLCK VI_A group are the only known plant kinases for which the activity is directly regulated by ROP GTPases, at least in vitro [6,8,9]. This fact is rather surprising, considering the wide role of animal Rho-type GTPases in kinase signalling [10], as well as the central role of ROP GTPases in a variety of cellular functions [13]. Despite their unique regulation, the biological function of RLCK VI_A kinases has hardly been investigated so far [11]. The barley HvRBK1 kinase (homologue of the Arabidopsis RLCK VI_A3 kinase) was shown to have a role in basal disease resistance [6]. Transient silencing of the gene decreased the stability of cortical microtubules and promoted fungal penetration into barley epidermal cells. Mutation in the gene coding for the Arabidopsis homologue of HvRBK1, AtRLCK VI_A3 was reported to support fungal reproduction [9]. Arabidopsis *AtRBK1* (*RLCK VI_A4*) and *AtRBK2* (*RLCK VI_A6*) genes were shown to have augmented expression following pathogen infection, supporting the general role of RLCK VI_A kinases in pathogen responses. The *atrlck vi_a3* mutant also exhibited reduced plant size and an increase in the ratio of trichomes with high branch numbers, while the AtRBK1 (RLCK VI_A4) kinase was found to be a member of a kinase cascade regulating auxin-mediated cell elongation, consistent with a developmental/morphogenic role [14].

Here, we report the involvement of the *AtRLCK VI_A2* gene in the regulation of plant growth and (skoto)morphogenesis. T-DNA insertion into the gene resulted in reduced hypocotyl elongation and smaller rosette size, which could be ascribed to limited cell expansion. These mutant phenotypes were complemented by the exogenous application of gibberellic acid. Measurements could not reveal differences neither in the gibberellin content nor in the gibberellin sensitivity of the mutant. Transcript analysis indicated that the kinase might indirectly affect gibberellin-dependent responses during skotomorphogensis, by overlapping with the action of the central transcription factor network regulating hypocotyl growth, and interfering with hormone signalling, cell wall organisation, and cellular transport processes.

2. Results

2.1. Molecular Characterization of the RLCK VI_A2 T-DNA Insertion Mutant and the Transgenic Plants Used in the Study

In order to reveal possible biological functions of the ROP GTP-ase binding AtRLCK VI_A2 (At2G18890) kinase [8], we carried out a search in the GABI-Kat Arabidopsis T-DNA insertional mutant collection [15]. Two lines with predicted T-DNA insertion in the 5′ untranslated region (GABI_676D12) or in the second intron of the At2G18890 gene (GABI_435H03), respectively, could be identified. Seeds of the T-DNA mutants were obtained from the Nottingham Arabidopsis Stock Centre (NASC) [16]. Expression of the At2G18890 gene was tested in the homozygous lines using a specific PCR primer pair, amplifying the whole transcript by RT-PCR. It could be established that the GABI_435H03 line does not produce *RLCK VI_A2* transcripts in contrast to the line GABI_676D12 (Figure 1a). In order to determine the exact T-DNA insertion site, the junction region of the At2G18890 gene and the T-DNA was amplified from the genomic DNA of the GABI_435H03 line using a T-DNA-specific reverse primer and an At2G18890 second exon specific forward PCR primer. Sequencing of the PCR product verified that the T-DNA insertion was located not in the second intron but in the third exon (at the ninth codon after the intron/exon junction) (Figure 1b). Mapping transcript reads of the *rlck vi_a2* mutant by next generation sequencing (NGS) to the reference *Arabidopsis thaliana* genome confirmed that, although the first two exons are transcribed in the mutant, full length functional transcripts are not produced and therefore the mutant can be considered as a knock out (Supplementary Figure S1).

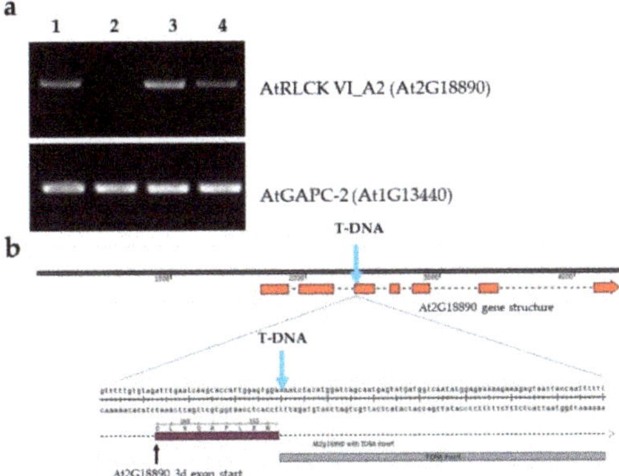

Figure 1. Expression of the receptor-like cytoplasmic kinase (*RLCK*) *VI_A2* gene in various Arabidopsis lines used in the study. (**a**) RT-PCR results using *RLCK VI_A2* (upper row) and *GAPC-2* specific (lower row) primers in wild type (**1**), T-DNA insertion line GABI_435H03 (**2**), T-DNA insertion line GABI_676D12 (**3**), and the GABI_435H03 line expressing the *RLCK VI_A2* cDNA transgene under the control of the 35S promoter (complemented mutant) (**4**). (**b**) Site of the T-DNA insertion in the third exon of the At2G18890 gene coding for the AtRLCK VI_A2 kinase in the GABI_435H03 line.

To validate the mutant phenotypes, various transgenic Arabidopsis lines were produced. The *rlck vi_a2* mutant was complemented with 35S-promoter-driven expression of the At2G18890 cDNA N-terminally fused, with a TAP-tag that allows co-immunoprecipitation of kinase interacting proteins [17]. Based on gene expression verification, a representative line was selected for further studies (Figure 1a). Furthermore, estradiol-induced RNA-interference [18] was used to knock down *RLCK VI_A2* expression (Supplementary Figure S2a) to further verify some of the experimental findings obtained with the mutant line.

2.2. The RLCK VI_A2 Kinase Controls Seedling and Plant Growth

The *rlck vi_a2* mutant seedlings that were grown under 8 h/16 h light/dark periods for 5 days exhibited significantly shorter hypocotyls as compared to the wild type (Figure 2a,c). Ectopic expression of the kinase in the mutant background restored hypocotyl growth to normal (Figure 2a,c). The cotyledons were also smaller in the mutant, although this difference was not statistically significant (Figure 2b). Since hypocotyl growth is accelerated in the dark, seeds were also germinated and cultured under continuous darkness for 16 days, and the size of the hypocotyls and cotyledons was compared (Figure 2d–f).

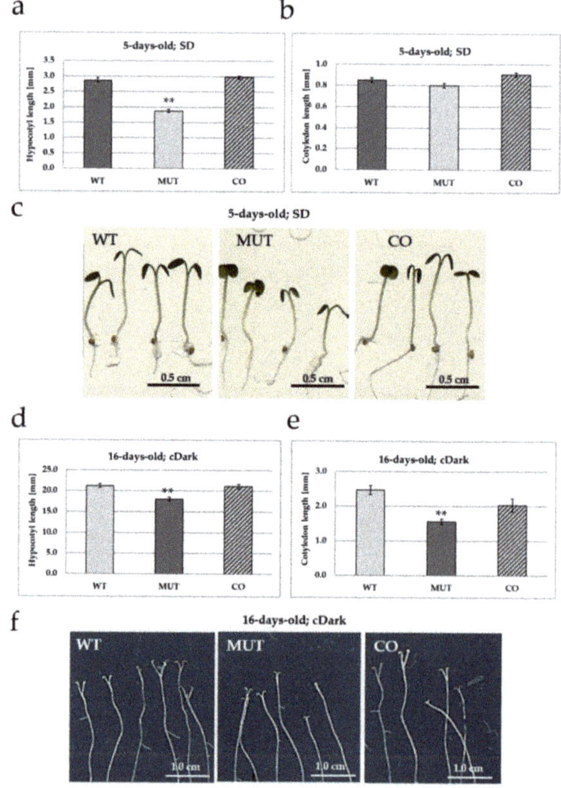

Figure 2. The *rlck vi_a2* mutation affects hypocotyl and cotyledon elongation. Hypocotyl (**a,d**) and cotyledon (**b,e**) length were measured for 5-days-old short-day (SD; 8/16h light/dark cycle) and 16-days-old dark-grown (cDark; continuous dark) seedlings. WT—wild type; MUT—T-DNA insertion mutant line; CO—complemented mutant line. Three biological replicates were made with 15–25 plants per line. Averages and standard errors are shown. Corresponding representative images are displayed on (**c,f**). ** $p < 0.005$ (Student's *t*-test; comparison to WT).

The hypocotyl of the mutant was found to be significantly shorter under this condition as well, while the wild type and complemented lines exhibited similar hypocotyl sizes (Figure 2d,f). The length of cotyledons, especially that of their petioles, were significantly reduced in the mutant but was restored to the wild type level in the complemented line (Figure 2e,f). The phenotypes of mutant seedlings could be recreated via estradiol-induced silencing of the *RLCK VI_A2* gene in transgenic seedlings (Supplementary Figure S2b).

Seedlings were also grown into plants in pots in the greenhouse under short-day condition (8 h light, 16 h dark). A significant difference in the size of the rosettes was observed: it was decreased in the mutant, but restored to the normal level in the complemented transgenic line as compared to the wild type control (Figure 3).

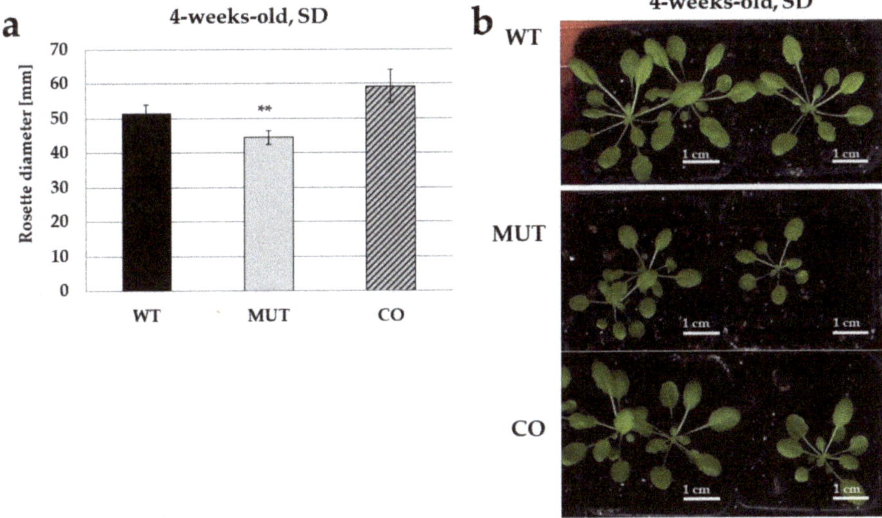

Figure 3. Greenhouse-grown mutant (MUT) plants exhibited smaller plant size, as evidenced by measuring the rosette diameter. Normal size of the wild type (WT) plants was restored by expressing the kinase cDNA in the mutant background (complemented, CO line). Rosette diameters in mm are shown in (**a**), and representative images of the measured 4-weeks-old plants in (**b**). The plants were grown in short day conditions (SD). Averages and standard errors were calculated and are shown on (**a**). $n = 15$–25, ** $p < 0.005$ (Student's *t*-test; comparison to WT).

Altogether, these data indicate that the RLCK VI_A2 kinase is required for normal plant growth under light as well as in dark; in seedlings as well as in greenhouse plants.

2.3. The RLCK VI_A2 Kinase Controls Cell Size

Microscopic investigations revealed that epidermal cell size was significantly smaller in both investigated organs of the mutant seedlings (Figure 4). In the mutant, hypocotyl epidermal cells were less elongated (Figure 4a,b), while the epidermal cells of the cotyledon were not only smaller in area, but their shape was also different; their circularity index was much higher (Figure 4c,d), indicating limited planar polarity [19].

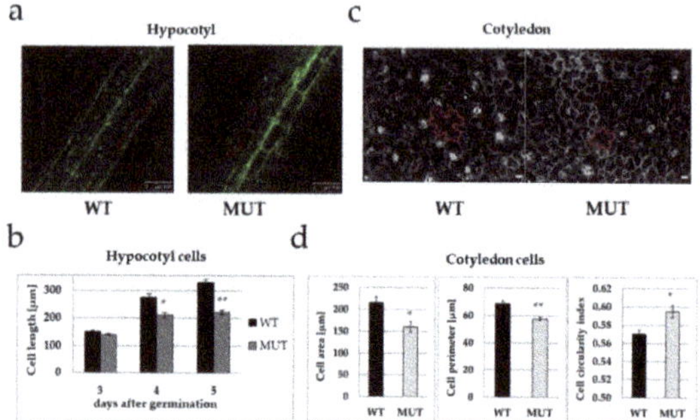

Figure 4. The sizes of hypocotyl (**a**,**b**) and cotyledon (**c**,**d**) cells are significantly smaller in the mutant (MUT) than in the wild type (WT) plants. Fluorescent (**a**) and scanning electron microscopic (**c**) images are shown for the epidermal cells of the hypocotyl (**a**) and the cotyledon (**c**), respectively, of 5-day-old seedlings. The white bars indicate 100 μm (**a**), and 10 μm (**b**), respectively. For the quantitative comparison of cell size (**b**,**d**), 150–200 cells were measured for each of three randomly selected seedlings per line. Averages and standard errors are shown on the histograms. * $p < 0.05$; ** $p < 0.005$ (Student's *t*-test; comparison to WT).

2.4. Gibberellic Acid Treatment Rectifies the rlck vi_a2 Mutant Phenotypes

In order to test whether the mutant phenotype can be linked to the disturbed action of plant hormones, seedlings were grown in the presence of various plant growth regulators (5 nM indoleacetic acid, 1 μM brassinolide, 100 μM ethephon (Ethrel) or 20 μM gibberellic acid (GA₃)) and hypocotyl, cotyledon and rosette sizes were measured (Supplementary Figures S3 and S4). Of the investigated hormones, exogenous gibberellin (GA₃) was found to rectify the growth defects of the *rlck vi_a2* mutant (Figure 5) and the RNAi-silenced lines (Supplementary Figure S4).

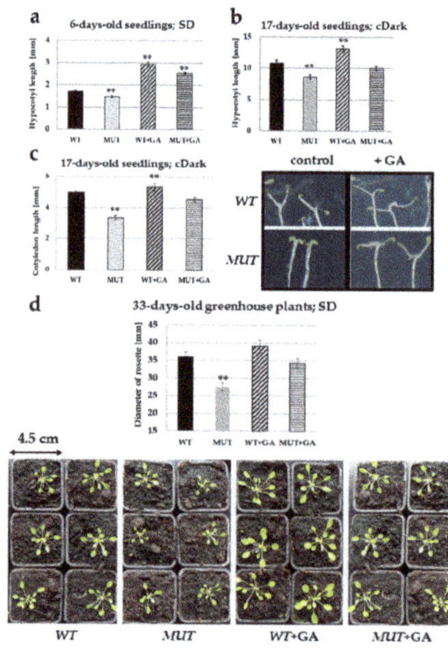

Figure 5. Exogenous gibberellin treatments complemented the mutant phenotypes. Wild type (WT) or *rlck vi_a2* mutant (MUT) seedlings grown in vitro in short days (SD; 8 h/16 h light/dark) for 6 days (**a**)

or in continuous dark (cDark) for 17 days (**b,c**) and plants grown at short days in greenhouse (**d**) were or were not treated with 20 μM gibberellic acid (GA$_3$). Hypocotyl length (**a,b**), cotyledon length (**c**) or rosette diameter (**d**) were measured in 15–25 seedlings or plants, respectively, in three repetitions. Averages and standard errors are shown. ** $p < 0.05$ (Student's *t*-test; comparison to WT).

2.5. The Effect of the rlck vi_a2 Mutation on Gibberellic Acid Level, Synthesis, and Signalling in Seedlings

Endogenous level of gibberellins (GAs) having biological activity was determined in the case of wild type, and mutant and complemented mutant seedlings (Figure 6a). The major bioactive GAs include GA$_1$, GA$_3$, GA$_4$, and GA$_7$, but GA$_5$ and GA$_6$ have also been indicated to have biological activities [20]. In 9-day-old seedlings, among these GAs, GA$_5$ exhibited the highest concentration, GA$_1$, GA$_4$ and GA$_7$ was found to have lower levels, while GA$_3$ and GA$_6$ could not be detected. No significant differences were found in the concentration of active GAs in the seedlings of the various Arabidopsis lines tested. The GA sensitivity of the mutant was also compared to that of the wild type control. It was found that hypocotyl growth was not less responsive to exogenous GA$_3$ in the mutant than in the wild type (Figure 6b).

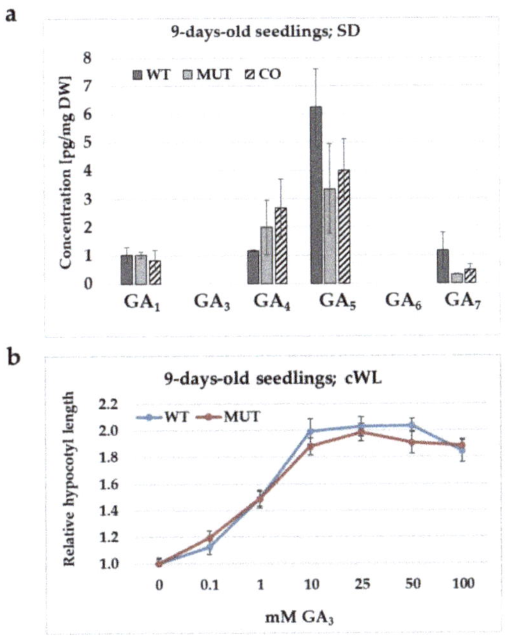

Figure 6. Gibberellin content and gibberellin sensitivity of the mutant and the wild type. (**a**) Endogenous content of active gibberellins was measured in 9-days-old seedlings of wild type (WT), *rlck vi_a2* mutant (MUT), and complemented mutant (CO) lines. Seedlings were grown in vitro under short day (SD; 8 h/16 h light/dark) conditions in a growth chamber. Samples were collected from three independent experiments. Averaged data are shown with the standard deviations. No statistically significant differences could be observed among the tested Arabidopsis lines ($p < 0.05$ Student's *t*-test; comparison to WT). (**b**) Relative hypocotyl length was determined in response to a range of GA$_3$ concentrations (0–100 μM), in the case of wild type and mutant seedlings (9-days-old; grown under low intensity continuous white light).

2.6. Transcriptome Analysis Indicated a Gibberellin-Independent Role for the RLCK VI_A2 Kinase in Skotomorphogenesis

To clarify the role of the RLCK VI_A2 kinase in seedling growth and cell elongation, the transcriptome of 18-days-old dark-grown wild-type and *rlck vi_a2* mutant seedlings were compared in three biological replicates. 1203 statistically highly significant ($q < 0.05$; fold change > 1.5) DEGs (differentially expressed genes) were identified, among which 406 were upregulated and 707 downregulated (Supplementary Table S1). Gene ontology (GO) enrichment analysis (Plant GOSlim ontologies) indicated the involvement of the kinase in the responses to various exogenous and endogenous stimuli including biotic and abiotic stresses and plant hormones as well as in lipid, carbohydrate, and secondary metabolism (Table 1). A more detailed GO analysis (complete GO ontologies) showed that the kinase might modulate the signalling, metabolism and transport of several hormones, including auxin, abscisic acid, jasmonic acid, and salicylic acid, but gibberellic acid-related DEGs were not significantly enriched in the mutant (Supplementary Tables S2 and S3).

Table 1. Gene ontology classification (Plant GOSlim) of the DEGs for 17-days-old dark-grown *rlck vi_a2* mutant seedlings without roots in comparison to wild type.

GO Term	Ontology	Description	p-Value	FDR
GO:0050896	P	response to stimulus	8.6×10^{-66}	2.7×10^{-63}
GO:0006950	P	response to stress	1.4×10^{-47}	2.2×10^{-45}
GO:0009605	P	response to external stimulus	1.8×10^{-36}	2.00×10^{-34}
GO:0009607	P	response to biotic stimulus	1.9×10^{-29}	1.6×10^{-27}
GO:0009628	P	response to abiotic stimulus	1.2×10^{-27}	7.4×10^{-26}
GO:0051704	P	multi-organism process	2.2×10^{-24}	1.2×10^{-22}
GO:0009719	P	response to endogenous stimulus	8.1×10^{-24}	3.7×10^{-22}
GO:0019748	P	secondary metabolic process	2.6×10^{-20}	1.00×10^{-18}
GO:0007154	P	cell communication	3.7×10^{-15}	1.3×10^{-13}
GO:0009056	P	catabolic process	6.3×10^{-12}	2.00×10^{-10}
GO:0007165	P	signal transduction	2.8×10^{-11}	8.3×10^{-10}
GO:0009987	P	cellular process	2.7×10^{-10}	7.2×10^{-9}
GO:0008152	P	metabolic process	2.2×10^{-9}	5.4×10^{-8}
GO:0051179	P	localization	3.3×10^{-9}	7.6×10^{-8}
GO:0009991	P	response to extracellular stimulus	1.5×10^{-8}	3.1×10^{-7}
GO:0006629	P	lipid metabolic process	2.1×10^{-8}	4.1×10^{-7}
GO:0006810	P	transport	3.2×10^{-8}	6.1×10^{-7}
GO:0051234	P	establishment of localization	4.7×10^{-8}	8.4×10^{-7}
GO:0005975	P	carbohydrate metabolic process	2.2×10^{-7}	3.7×10^{-6}
GO:0008219	P	cell death	7.9×10^{-6}	0.00013
GO:0065008	P	regulation of biological quality	2.00×10^{-5}	0.00031
GO:0065007	P	biological regulation	2.7×10^{-5}	0.00039
GO:0042592	P	homeostatic process	5.6×10^{-5}	0.00078
GO:0040007	P	growth	0.0009	0.012
GO:0009606	P	tropism	0.0011	0.014
GO:0032502	P	developmental process	0.0017	0.021
GO:0050789	P	regulation of biological process	0.002	0.024
GO:0050794	P	regulation of cellular process	0.0025	0.029
GO:0048856	P	anatomical structure development	0.0038	0.042
GO:0003824	F	catalytic activity	1.3×10^{-25}	1.3×10^{-23}
GO:0005215	F	transporter activity	4.2×10^{-9}	2.1×10^{-7}
GO:0016740	F	transferase activity	1.9×10^{-8}	6.4×10^{-7}
GO:0008289	F	lipid binding	1.2×10^{-7}	3.00×10^{-6}
GO:0030246	F	carbohydrate binding	5.7×10^{-6}	0.00012
GO:0016787	F	hydrolase activity	1.2×10^{-5}	0.0002
GO:0019825	F	oxygen binding	1.9×10^{-5}	0.00027
GO:0016301	F	kinase activity	2.5×10^{-5}	0.00032
GO:0000166	F	nucleotide binding	0.00014	0.0016

Table 1. *Cont.*

GO Term	Ontology	Description	p-Value	FDR
GO:0005488	F	binding	0.00021	0.0021
GO:0060089	F	molecular transducer activity	0.00041	0.0034
GO:0004872	F	receptor activity	0.00041	0.0034
GO:0016772	F	transferase activity, transferring phosphorus-containing groups	0.0012	0.009
GO:0005576	C	extracellular region	8.6×10^{-37}	1.6×10^{-34}
GO:0030312	C	external encapsulating structure	6.7×10^{-34}	4.2×10^{-32}
GO:0005618	C	cell wall	6.7×10^{-34}	4.2×10^{-32}
GO:0005886	C	plasma membrane	7.5×10^{-32}	3.5×10^{-30}
GO:0016020	C	membrane	1.3×10^{-25}	4.9×10^{-24}
GO:0005773	C	vacuole	4.1×10^{-5}	0.0013
GO:0005783	C	endoplasmic reticulum	8.8×10^{-5}	0.0023
GO:0012505	C	endomembrane system	0.0018	0.043

FDR—false discovery rate; P—biological process; F—molecular function; C—cellular compartment.

Browsing the annotation of those DEGs that have the GO "hormone response" for the key word "gibberellin" revealed a small number of GA-regulated proteins (Table 2 and Supplementary Table S3), including the GAI DELLA-type transcriptional regulator, which exhibited an app. 1,5-fold increased expression in the mutant. The expression of other DELLA protein genes showed a slightly but not statistically significant increase in this line. GA-related DEGs also included the AtHB23 homeobox protein that is involved in light-regulated hypocotyl growth and cotyledon expansion [21] and the GA catabolic enzyme GA2ox6 [22] (Table 2 and Supplementary Table S3). GA transport is considered to be mediated by NPF (NRT1/PTR FAMILY) transporters [23–25]. The NPF family has 53 members in Arabidopsis [26], but the expression of only 6 of them was altered in the *rlck vi_a2* mutant. Interestingly, four out of these 6 NPFs have already been implicated in GA transport (Table 2 and Supplementary Table S3) [24,25,27]. Expressions of selected GA-related transcripts in the mutant and wild type background were also tested by qRT-PCR (Supplementary Figure S5a). The results supported the validity of the transcriptomic data.

Table 2. Manually identified DEGs implicated in gibberellin metabolism/response/transport in dark-grown 17-days-old *rlck vi_a2* mutant seedlings (see Supplementary Table S3 for more details).

Gene_ID	Annotation	Fold Change	q_Value	Significant
DEGs Implicated in GA Metabolism/Response *				
AT1G26960	AtHB23 homeobox protein 23	2.25	0.026242	yes
AT1G74670	GASA6 Gibberellin-regulated family protein	2.07	0.001941	yes
AT2G37640	EXP3 Barwin-like endoglucanases superfamily protein	1.95	0.001941	yes
AT5G15230	GASA4 GAST1 protein homolog 4	1.88	0.001941	yes
AT2G14900	AT2G14900 Gibberellin-regulated family protein	1.70	0.001941	yes
AT1G14920	GAI GRAS family transcription factor family protein	1.46	0.00859	yes
AT3G11280	AT3G11280 Duplicated homeodomain-like superfamily protein	0.69	0.00859	yes
AT4G19700	RING SBP (S-ribonuclease binding protein) family protein	0.68	0.038936	yes
AT1G75750	GASA1 GAST1 protein homolog 1	0.56	0.003506	yes
AT5G44610	MAP18 microtubule-associated protein 18	0.55	0.001941	yes
AT1G02400	GA2OX6 gibberellin 2-oxidase 6	0.58	0.044627	yes
DEGs of NPF (NRT1/PTR FAMILY) Gibberellin Transporters **				
AT1G52190	AT1G52190 Major facilitator superfamily protein, NPF1.2 [a,b]	1.60	0.001941	yes
AT3G16180	AT3G16180 Major facilitator superfamily protein, NPF 1.1 [a]	1.79	0.001941	yes
AT5G46050	PTR3 peptide transporter 3, NPF 5.2 [a]	0.55	0.001941	yes
AT5G62680	GTR2 Major facilitator superfamily protein, NPF 2.11 [c]	0.65	0.031474	yes
DEGs of DELLA Transcription Regulators ***				
AT1G14920	GAI GRAS family transcription factor family protein	1.46	0.00859	yes
AT2G01570	RGL1 RGA-like 1	1.25	0.483175	no
AT1G66350	RGA1 GRAS family transcription factor family protein	1.11	0.773976	no

Table 2. *Cont.*

Gene_ID	Annotation	Fold Change	*q*_Value	Significant
AT3G03450	RGL2 RGA-like 2	1.58	0.496428	no
AT5G17490	RGL3 RGA-like protein 3	1.21	0.819669	no

* Manually selected based on the presence of the word "gibberellin" in the annotations of DEGs implicated in "hormone response" by the AgriGO v2.0 tool. ** NPF transporters that were reported to transport gibberellins by [a] [24]; [b] [27]; [c] [25]. *** Note that only the differential expression of *GAI* was statistically significant.

Although only a few of them were directly GA-related, the *RLCK VI_A2*-dependent DEGs showed an overlap (15%), with the DEGs identified in the GA synthesis defective mutant *ga1-3* [28,29] (Figure 7 and Supplementary Table S4) suggesting an indirect and limited effect of the kinase on GA signalling.

Hypocotyl elongation is coordinated by the light and GA-regulated PIF transcription factors [30], as well as by the brassinosteroid- and auxin-controlled transcription factors BZR1 and ARF6, respectively [31]. The transcription factors were shown to interact with each other, forming a central growth regulatory circuit [31]. Promoters of thousands of genes were identified to be direct and partly common targets of the PIF4, BZR1 and ARF6 factors in relation to hypocotyl cell elongation [31]. Among the 1019 DEGs affected by the *rlck vi_a2* mutation, 317 (31%) belong to direct PIF4 targets, 359 (35%) is the direct target of BZR1 and 176 (17%) of ARF6, respectively, and 100 (9.8%) of them bind all three TFs (Figure 7 and Supplementary Table S5). PIF4 itself was found to be slightly upregulated (1.5-fold) in the mutant. Expressions of selected transcripts involved in the regulation of hypocotyl elongation during skotomorphogenesis were also tested in the mutant and the wild type by qRT-PCR (Supplementary Figure S5b). The obtained results supported the validity of the transcriptomic data.

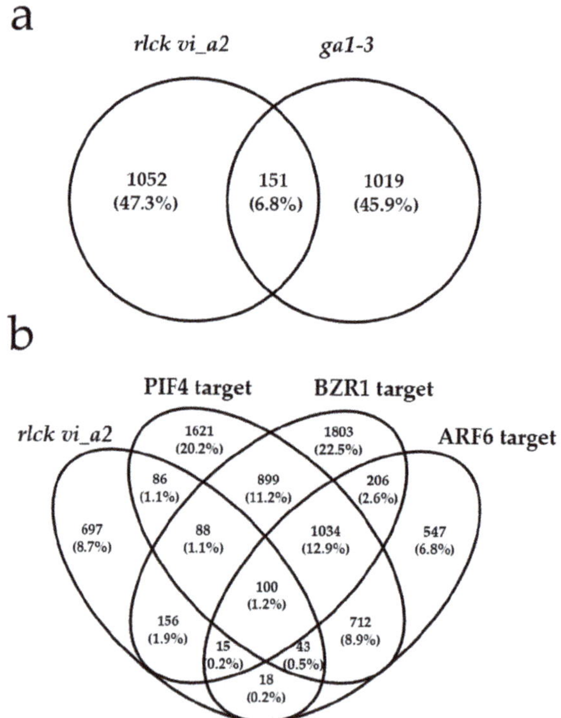

Figure 7. Overlaps of the DEGs of the *rlck vi_a2* mutant with the DEGs of the *ga1-3* gibberellin synthesis mutant [28,29] (**a**) and with those genes for which the promoters are direct targets of the cell/hypocotyl elongation regulatory transcription factors PIF4, BZR1, and/or ARF6 [31] (**b**).

Altogether, these data imply that GA level/signalling is not the primary target of the RLCK VI_A2 kinase, at least not at the gene transcription level.

2.7. Transcriptome Analysis Revealed the Role of the RLCK VI_A2 Kinase in Cellular Transport and Cell Wall Organisation

High number of the *RLCK VI_A2*-related DEGs code for proteins with catalytic, transport, transducer or binding activities, respectively (Table 1). Considering cellular localisation, proteins with extracellular (279, 23%) or cell periphery (395, 33%) localisation, and with association to cellular membranes (530, 44%), including the endomembrane system (118, 10%), were significantly overrepresented by the DEGs (Table 1 and Supplementary Tables S2 and S6). Moreover, 6% (72) of the DEGs are implicated in cell wall organisation and biogenesis (Supplementary Tables S2 and S6). Of note is the upregulation of several expansin and xyloglucan endotransglucosylase/hydrolase coding genes and the downregulation of those coding for extensin-like family proteins in the mutant background (Supplementary Table S6; for qRT-PCR validation Supplementary Figure S5c). Eight percent of DEGs (101) code for proteins that have transmembrane transport activities, including ion, nutrient or hormone transporters (Supplementary Table S6). Worth mentioning of the upregulation of several tonoplast intrinsic proteins and a number of auxin transporters, such as PIN4, PIN7, EIR1, ABCB19, LAX2, and the downregulation of many ion transporters (phosphate, sulphate, potassium etc.). These data indicate that RLCK VI_A2-mediated protein phosphorylation is required for the proper cellular transport of a wide variety of ions, nutrients, structural and regulatory molecules.

3. Discussion

Plant growth and development have to be continuously harmonized with external conditions. Protein kinases have central roles in sensing environmental signals, as well as in coordinating cellular and developmental responses. Plants possess a large superfamily of diverse protein kinase types, including the signal-sensing transmembrane receptor kinases, RLKs, and various types of downstream intracellular signal-transducing kinases. Among the latter, receptor-like cytoplasmic kinases (RLCKs) have recently gained increasing interest as potential mediators and modulators of RLK signalling [4,32,33]. Nevertheless, only a few of the 149 Arabidopsis and 187 rice RLCKs have been characterized and have known functions. Here, we describe the characterization of a T-DNA insertion mutant of At*RLCK VI_A2* that implies roles for the kinase in the regulation of cell/plant growth and morphogenesis.

3.1. The RLCK VI_A Kinases Are Required for Cell Elongation and Organ Growth in Addition to Their Role in Stress Responses

The RLCK VI_A3 kinase of Arabidopsis and its barley homologue, HvRBK1, have been implicated in ROP-GTPase-dependent pathogen resistance/susceptibility reactions [6,7,9], while the AtRLCK VI_A4/AtRBK1 and AtRLCK VI_A6/AtRBK2 kinases were shown to be expressed in response to pathogen infection [7], suggesting their primary role in plant defence. A considerable number of DEGs of the *rlck vi_a2* mutant are also related to plant defence (Table 1), strengthening this view. In addition, the transcript analysis indicated that the kinase might function during abiotic stress responses as well (Table 1). However, the involvement of several RLCK VI_A kinases in the regulation of plant growth and development has also been reported: the T-DNA mutant of the related *AtRLCK VI_A3* kinase is stunted and has over-branched trichomes [9]; the Arabidopsis RBK1 protein kinase (RLCK VI_A4) has been implicated in auxin-responsive cell expansion, due to the reduced auxin responsiveness of its T-DNA insertion mutant [14].

Our results show that the mutant seedlings producing no full length RLCK VI_A2 kinase (Figure 1) have limited cell expansion as compared to the wild type (Figure 3) resulting in shorter hypocotyls and cotyledons (Figure 2). Independent experiments using RNA interference to reduce *RLCK VI_A2* expression (Supplementary Figure S2), as well as the complementation of the T-DNA-caused mutation in transgenic lines (Figure 2) confirmed that the observed phenotypes are indeed associated with the absence of the kinase. Greenhouse-grown mutant plants also exhibited smaller rosette/leaf size than the wild type similarly to the reduced plant size of the related AtRLCK VI_A3 kinase [9].

Altogether, these observations strengthen the view that RLCK VI_A members have a general role in cell expansion and plant growth. Interestingly, while mutation in *AtRBK1/AtRLCK VI_A4* resulted in increased auxin sensitivity, in our experiments, the *atrlck vi_a2* mutant showed no altered auxin response (Supplementary Figure S3), but the mutant phenotypes could be rescued by exogenous GA_3 (Figure 5 and Supplementary Figure S4). This indicates that the various RLCK VI_A kinases might influence cell expansion via various pathways.

3.2. How the RLCK VI_A2 Kinase May Affect Cell Expansion?

How RLCK VI_A2 kinase regulate cell expansion is not known at present. It has to be mentioned that the barley HvRBK1/HvRLCK VI_A3 kinase has been shown to be required for proper cortical microtubule organisation; the silencing of HvRBK1 was shown to result in a fragmented cortical microtubule network [6]. Since cortical microtubules are known to control directional cell elongation [34,35] in a ROP GTPase-dependent manner [36], the ROP-binding kinases might be involved in this process. Transcriptome analysis during the skotomorphogenesis of *rlck vi_a2* mutant seedlings indicates that high portion of the DEGs modulated in the mutant code for proteins located at membranes, at the cell periphery or in the apoplast, and may have a role in transport or cell wall organisation (Table 1 and Supplementary Tables S2 and S6). Therefore, it is conceivable that the RLCK VI_A2 kinase has a regulatory role in these processes in relation to cell elongation. The transcriptomic data, however, do not provide a clear view about the role of the kinase in cell elongation. The upregulation of several expansins and xyloglucan endotransglucosylase/hydrolase coding genes implicated in cell wall loosening [37,38] and the downregulation of those coding for extensin-like family proteins rather contributing to cell wall stiffening [39,40] are not consistent with the observed phenotype of the mutant having restricted cell elongation. Moreover, PIF4, a positive regulator of skotomorphogenesis including hypocotyl cell elongation is upregulated in the *rlck vi_a2* mutant, despite its short-hypocotyl phenotype [30]. These contradictions might be resolved by keeping in mind that the kinase primarily modulates posttranslational and not transcriptional regulation. The observed transcriptional changes might be indirect responses to the missing kinase function: blocking cell elongation at the posttranslational level (e.g., phosphorylation-dependent degradation of PIF4 or other regulators) might give a feedback to increase the transcription of genes promoting cell expansion. For example, PIF4 has been shown to be phosphorylated by the brassinosteroid signalling kinase BRASSINOSTEROIDINSENSITIVE 2 (BIN2), marking it for proteasomal degradation [41]. Considering the number of genes affected by the *rlck vi_a2* mutation, and being at the same time the direct targets of the PIF4/BZR1/ARF6 transcriptional factor circuit that centrally controls cell expansion and hypocotyl growth (Figure 7 and Supplementary Table S5), one can suppose that the kinase directly and/or indirectly modulates the downstream processes controlled by these factors.

3.3. Gibberellin might Indirectly Complement for the Missing Kinase Function

Although exogenous GA_3 treatment could rectify the absence of RLCK VI_A2 function (Figure 5 and Supplementary Figure S4), the kinase mutant exhibited similar bioactive GA levels than the wild type (Figure 6a). Moreover, there was no decrease in the GA sensitivity of the mutant (Figure 6b) despite a ca. 1.5-fold increase in the expression of the GRAS-domain GAI protein gene (Table 2, Supplementary Figures S3 and S5a), a negative regulator of GA signalling. Transcriptomic analysis confirmed that genes implicated in gibberellin metabolism are hardly affected by *RLCK VI_A2* expression. However, the same analysis indicated the misregulation of 15% of *RLCK VI_A2*-dependent DEGs also in the GA synthesis defective mutant *ga1-3* [28,29] (Figure 7a and Supplementary Table S4). Furthermore, the RLCK VI_A2 kinase might be involved in the modulation of GA transport, since the expression of four potential GA transporters [25,26,28] was found to be regulated in the mutant background (Table 2).

The above observations indicate that, although RLCK VI_A2 functions interfere with GA action, this is likely not through the direct modulation of the synthesis of bioactive GAs. Transcriptomic analysis revealed that the absence of the RLCK VI_A2 kinase affected the signalling of auxin and BR, the two other hormones also centrally involved in the light and developmental regulation of hypocotyl elongation [31]. Both the auxin and BR hormones are well known to crosstalk with GA modulating each other's metabolism and signalling [37–39]. Although all three hormones (GA, BR and auxin) act on distinct TFs governing hypocotyl elongation (PIF4, ARF6, and BZR1, respectively), the target genes of these transcription factors largely overlap [31] (Figure 7b). A considerable fraction of the DEGs of the *rlck vi_a2* mutant (31.6%) are direct targets of at least one of the above TFs, while 10% of the DEGs is direct target of all three. The observed gene expression changes might be indirect consequences of blocked cell expansion/seedling growth. The data support the view that the RLCK VI_A2 kinase might control basic cell elongation processes downstream of the PIF4/BZR1/ARF6 TFs, rather than the regulatory proteins themselves.

Why exogenous GA_3, but not auxin or brassinosteroid, rescue the *rlckvi_a2* mutant phenotypes is not known. Exogenous GA_3 might induce parallel pathways that can overcome the cell elongation defects caused by missing protein phosphorylations in the absence of the RLCK VI_A2 kinase (e.g., via other kinases and/or microtubule/protein stability and/or cell wall organisation, etc.). One of such shared pathways is the regulation of auxin transport, since several auxin transport protein genes are regulated in the mutant background (Supplementary Table S3), and GA is known to affect auxin transport stabilizing these proteins [42].

4. Materials and Methods

4.1. Plant Material and Growth Conditions

Seeds of *Arabidopsis thaliana* (L.) Columbia-0 and mutant lines of the GABI-Kat Arabidopsis T-DNA insertional mutant collection [15], GABI_435H03 and GABI_676D12, were obtained from the Nottingham Arabidopsis Stock Centre (NASC).

The full-length cDNA of At2G18890 was obtained from the Arabidopsis Biological Resource Centre (ABRC, Columbus, OH, USA; stock number U67191). The cDNA was amplified (denaturation 94 °C 10 s, annealing temperature 62 °C for 30 s, elongation 1 min at 72 °C) by the proof-reading PHUSION™ II polymerase (Thermo Fisher Scientific, Waltham, MA, USA), using specific primers, having added EcoRI and XhoI sites (Supplementary Table S7). The gel-purified PCR fragment was digested by FastDigest™ EcoRI and XhoI enzymes (Thermo Fisher Scientific) and inserted into similarly cut and purified pENTR2B vector (Thermo Fisher Scientific). The cDNA was cloned into the plant expression vectors pN-TAPa (Gene bank accession: AY788908 [17]) and pMDC7 [18,43] via Gateway recombination, using standard LR Clonase™ II (Thermo Fisher Scientific, catalogue number: 11791019) reaction, as recommended by the supplier. The binary vectors were transformed into GV3101/pMP90 Agrobacterium strain with tri-parental mating [44], which were used for transgenic *Arabidopsis thaliana* (GABI_435H03 and/or Col-0) production via floral dip agroinfiltration [45]. Seeds were selected using appropriate antibiotics (gentamycin and hygromycin, respectively). Plants (including the T-DNA insertion mutants) were characterized for *RLCK VI_A2* (AT2G18890) expression, using the same primers as for cloning in reverse transcription polymerase chain rection (RT-PCR) (see later). Transgenic lines with appropriate expression were selfed and propagated. Stable, homozygous T3/T4 transgenic plants were used in the experiments.

Seeds of the wild type, mutant and transgenic lines were sterilized in 2% bleach, resuspended in sterile water and stratified (4 °C for 48 h). Germination was performed in vertically oriented square Petri dishes with half strength Murashige and Skoog (MS) medium containing 0.5% sucrose 0.8% agar, pH 5.7 (Duchefa Biochemie, Haarlem, The Netherlands). When estradiol inducible lines were used, the growth medium contained 5 µM β estradiol (E2758, Sigma, St. Louis, Mo, USA) for gene expression induction. The experiments were done in growth chambers (Aralab, Rio de Mouro, Portugal) under short days (8 h 22 °C, 120 µE light intensity and 16 h 21 °C darkness); in complete darkness at 22 °C to investigate skotomorphogenesis; or in continuous low white light at 60 µE for the gibberellin sensitivity assay.

4.2. Analysis of Hypocotyl Length and Rosette Size Measurement

The hypocotyl and cotyledon lengths were measured from digital photographs by the ImageJ software (NIH, Bethesda, MD, USA). At least 60 wild-type and mutant seedlings were analysed in three biological replicates. For the determination of plant rosette size, 28 or 33-days-old Arabidopsis plants grown in pots were photographed, and rosette diameters were measured with the "straight line" function of the ImageJ software [46]. The distance between the tips of the two longest rosette leaves were measured and exported to MS Excel file for further analysis. Images were taken with a digital camera (Fuji FinePix S1000fd) using the same parameters (focus distance, resolution, ISO).

4.3. Cell Size and Shape Analysis

The analysis of hypocotyl cell length was done after Acridine orange (100 µg/mL) staining. Images on the three different part of the hypocotyls (basal, middle, top) were taken by confocal laser scanning microscopy (TCS SP5, Leica Microsystems, Heidelberg, Germany), and the cell length was determined using the ImageJ software [46].

Cotyledon epidermal cell size and shape were visualized uncoated in a JSM-7100F/LV scanning electron microscope (JEOL Ltd., Akishima, Tokyo, Japan) at low-vacuum, by detecting backscattered electrons according to [47]. The images were taken from the same zone of the cotyledons (at the centre, next to the main vein) and perimeter, area, and circularity of cells were analysed using the ImageJ software [46].

Altogether, 150–200 cells were measured per line, in three repetitions.

4.4. Gibberellin Content Measurement by Ultra-High Performance Liquid Chromatography-Tandem Mass Spectrometry

The sample preparation and analysis of GAs were performed according to the method described in [48] with some modifications. Briefly, tissue samples of 26–60 mg DW (three independent technical replicates of each of the tree biological samples) were ground to fine consistency using 3-mm zirconium oxide beads (Retsch GmbH & Co. KG, Haan, Germany) and an MM 301 vibration mill at a frequency of 30 Hz for 3 min (Retsch GmbH & Co. KG, Haan, Germany), with 1 mL of ice-cold 80% acetonitrile, containing 5% formic acid as extraction solution. The samples were then extracted overnight at 4 °C using a benchtop laboratory rotator Stuart SB3 (Bibby Scientific Ltd., Staffordshire, UK), after adding 17 internal GA standards ([2H2]GA1, [2H2] GA3, [2H2]GA4, [2H2]GA5, [2H2]GA6, [2H2]GA7) purchased from OlChemIm, Olomouc, Czech Republic. The homogenates were centrifuged at $36,670 \times g$ and 4 °C for 10 min; corresponding supernatants further purified using reversed-phase and mixed mode SPE cartridges (Waters, Milford, MA, USA) and analysed by ultra-high performance liquid chromatography-tandem mass spectrometry (UHPLC-MS/MS; Micromass, Manchester, UK). GAs were detected using multiple-reaction monitoring mode of the transition of the ion [M–H]- to the appropriate product ion. Masslynx 4.1 software (Waters, Milford, MA, USA) was used to analyse the data and the standard isotope dilution method [49] was used to quantify the GAs levels.

4.5. Hormone Treatments

For hypocotyl and cotyledon measurements, the plants were germinated in vertically oriented square Petri dishes in 22 °C, under SD conditions or in continuous darkness as indicated. 6-day-old seedlings were moved to 5 nM IAA (I2886 Sigma, St. Louis, MO, USA)-containing medium for additional 6 days before measured. Epibrassinolid (1 µM; E1641 Sigma) or Ethrel (100 µM; Bayer CropSience, Gent, Belgium) or GA3 (20 µM; G7645 Sigma) were included into the medium from the beginning of culture. Hormone concentrations not inhibiting hypocotyl elongation were selected based on previous studies [50–53]. Transgenic RNAi seedlings and corresponding controls were grown in the presence of 5 µM β estradiol in addition to the hormones.

For the gibberellic acid sensitivity assay, constant 60 µE white light was used to limit dark-induced elongation at 22 °C. GA_3 was included into the growth medium in concentrations indicated on the figure.

For the complementation of the rosette size, 14-day-old, soil-grown plants were sprayed with 20 µM GA_3 solution supplemented with 0.01% Silwet L-77 (Kwizda, Vienna, Austria). The treatment was repeated at 4-day intervals. The control plants were sprayed with 0.01% Silwet L-77 solution. The GA_3 was dissolved in DMSO:methanol solution (1:1) and stock solutions were prepared at 1 µM (GA_3) concentrations for further dilution in water. Rosette size was determined as described earlier, 19 days following the start of the treatment (33-day-old plants).

4.6. Characterization of Mutant/Transgenic Plants by RT-PCR

The Quick-RNA Plant Miniprep Kit (Zymo Research, Irvine, CA, USA) was used to isolate total RNA from whole seedlings. Total RNA was treated by RNase-free DNase I (Thermo Fisher Scientific) and cDNA templates were generated from 0.5 mg RNA samples by RevertAid M-MuLV reverse transcriptase (Thermo Fisher Scientific). The full length transcript of the *RLCK VI_A2* gene was amplified with primers that were planned for cloning the cDNA in standard PCR reaction (denaturation 94 °C for 30 s, annealing temperature 55 °C for 30 s, elongation 1 min at 72 °C) with DreamTaq polymerase (Thermo Fisher Scientific). *AtGAPC-2* (AT1G13440) transcripts were used as internal reference. See primers in Supplementary Table S7.

Genomic DNA of the GABI_435H03 T-DNA insertion mutant was isolated with the Phire Plant Direct PCR Kit (Thermo Fisher Scientific) and the T-DNA insertion site was amplified with an At2G18890-specific forward primer (VIA2mid_F) and a T-DNA-specific reverse primer (T-DNA LB out), according to the supplier's instructions. The purified PCR products were sequenced using the same primers. For the primer sequences, see Supplementary Table S7.

4.7. RNA-Seq and Data Analysis

The Quick-RNA Plant Miniprep Kit (Zymo Research, Irvine, CA, USA) was used to isolate total RNA from 18-day-old dark-grown seedlings after removing their roots. The RNA preparations were quality checked and quantified using the Agilent RNA 6000 Nano Kit in an Agilent 2100 Bioanalyzer capillary gel electrophoresis instrument (Agilent, Santa Clara, CA, USA). For sequencing library preparation, polyA RNAs were selected from 800 ng total RNA using NEBNext Poly(A) mRNA Magnetic Isolation Module, then strand specific indexed libraries were prepared with NEBNext Ultra II Directional RNA Library Prep Kit for Illumina (New England Biolabs, Ipswich, MA, USA). Libraries were validated and quantified with an Agilent DNA 1000 kit in a 2100 Bioanalyzer instrument, then after pooling and denaturing, library pools were sequenced in an Illumina MiSeq instrument with MiSeq Reagent Kit V3-150 (Illumina Inc., San Diego, CA, USA), generating 2×75 bp paired-end reads. Fasq files were trimmed and adapter sequences removed with Trimmomatic 0.33 in paired-end mode [54]. Paired sequences were aligned to the TAIR10 Arabidopsis reference genome using TopHat2 [55]. Binary alignment (*.bam) files were sorted and deduplicated with SAMtools (http://samtools.sourceforge.net/), then differential expression analysis was done with Cufflinks (http://cufflinks.cbcb.umd.edu/), using Araport 11 transcript annotation [56]. Differential expression was considered as significant with a q value lower than 0.05.

The RNA-seq data used for the analysis have been deposited in the NCBI Sequence Read Archive (SRA) (https://www.ncbi.nlm.nih.gov/sra) under the accession PRJNA644816.

4.8. Real-Time Quantitative PCR (qRT PCR)

For qRT PCR, total RNA was purified and converted to cDNA as described under 4.6. The oligonucleotide primers are listed in Supplemental Table S7. A few of them have been previously published in [53]. qRT-PCR reactions were performed using an ABI PRISM 7700 sequence detection system (Thermo Fisher Scientific) and the qPCRBIO SyGreen Mix Hi-ROX master mix (PCR Biosystems Ltd., London, UK) using standard protocol (denaturation 95 °C for 10 min, 40 cycles of 95 °C for 10 s, and 62 °C for 60 s). Ct values were analysed using the RQ manager software (Thermo Fisher Scientific), and then exported to Microsoft Excel for further analysis. The ratio of each mRNA relative to the mRNA of the *Arabidopsis thaliana UBIQUITIN EXTENSION PROTEIN 1* gene (*UBQ1*, AT3G52590) was calculated using the $2^{-\Delta\Delta CT}$ method. *UBQ1* gene expression was uniform in the wild type and mutant background, as shown by the RNA-seq analysis (see in Supplementary Table S1). The average of the three technical repeats of the WT control was used as reference (unit 1) to calculate relative expression for each gene in the mutant background.

4.9. Statistical Analysis

Plant culture experiments were carried out in three independent replicates. The number of investigated individuals per replicate is given in each figure legend. Averages with standard errors are shown in the histograms for growth parameters having high and variable sample numbers (e.g., cell, hypocotyl and cotyledon length measurements). In qRT-PCR experiments, two independent biological samples each, with three technical replicates, were amalgamated and analysed together. Student's *t*-test was used for pairwise statistical comparison of the mutant/treated samples to the corresponding wild type/control ones (* indicates p-value < 0.05, ** indicates p-value < 0.005).

5. Conclusions

Decreased level or absence of the RLCK VI_A2 kinase in transgenic Arabidopsis lines resulted in restricted cell expansion and organ/plant size under short day conditions, as well as in continuous dark (skotomorphogenesis), in seedlings as well as in greenhouse plants, indicating the general role of the kinase in plant growth. Transcriptomic analysis confirmed that the kinase might be involved in the modulation of processes that are associated with cell membranes, and take place at the cell periphery or in the apoplast, such as cellular transport and cell wall organisation. Although exogenous GA$_3$ could rescue the mutant phenotypes, hardly any changes in gibberellin metabolism and/or signalling could be observed in the mutant, indicating that the RLCK VI_A2 kinase and gibberellin might act parallel on the same/similar processes. To clarify the exact role of the kinase in cell expansion and its hormonal regulation, the identification of its in vivo substrates is required.

Supplementary Materials: Supplementary materials can be found at http://www.mdpi.com/1422-0067/21/19/7266/s1. Figure S1: Mapping of transcript sequence reads to the genomic sequence of the At2G18890 gene, Figure S2: Estradiol-induced silencing of the *RLCK VI_A2* gene, Figure S3: Exogenous hormone treatments, Figure S4: GA$_3$ complements for the silencing of the *RLCK VI_A2* gene, Figure S5: qRT-PCR validation of the expression of selected genes, Table S1: Full transcriptomic analysis, Table S2: GO enrichment, Table S3: Hormone-responsive DEGs, Table S4: DEGs overlaping with the DEGs of the *gal-3* mutant, Table S5: DEGs directly regulated by the hypocotyl elongation controlling TFs, Table S6: DEGs related to cell wall-related and transport processes, Table S7: Sequences of the used oligonucleotide primers.

Author Contributions: Conceptualization, A.F. (Attila Fehér); formal analysis, L.B.; investigation, I.V., E.K., D.T. and A.F. (Anikó Faragó); methodology F.A. and I.D.; resources, A.F. (Attila Fehér); data curation, L.B.; writing—original draft preparation, A.F. (Attila Fehér); writing—review and editing, I.V.; supervision, A.F. (Attila Fehér), M.S. and L.B.; project administration, A.F. (Attila Fehér); funding acquisition, A.F. (Attila Fehér) and M.S. All authors have read and agreed to the published version of the manuscript.

Funding: This research including APC was supported by grants from the National Research, Development, and Innovation Office (NKFIH; #K124828) and the Hungarian Ministry for National Economy (GINOP-2.3.2-15-2016-00001). L.B. was supported by the János Bolyai Research Scholarship (BO/00522/19/8) of the Hungarian Academy of Sciences. The hormone measurements were supported from European Regional Development Fund Project "Centre for Experimental Plant Biology" (No. CZ.02.1.01/0.0/0.0/16_019/0000738) and the Czech Science Foundation (18-10349S).

Acknowledgments: The authors acknowledge the technical support provided by Róza Nagy and Hedvig Majzik (Institute of Plant Biology, Biological Research Centre).

Conflicts of Interest: The authors declare no conflict of interest. The funders had no role in the design of the study; in the collection, analyses, or interpretation of data; in the writing of the manuscript, or in the decision to publish the results.

Abbreviations

ABCB	ATP-binding cassette B protein
ARF	auxin response factor
AtHB	*Arabidopsis thaliana* homeobox-leucine zipper protein
AtRBK	*Arabidopsis thaliana* ROP-binding kinase
BZR	brassinazole-resistant
DEG	differentially expressed gene
DELLA	protein domain of the key negative regulators of GA action (DELLA proteins)
EIR	ethylene insensitivity of the root
FDR	false discovery rate
GA	gibberellin
ga1	gibberellic acid 1-gibberellin synthesis mutant
GAI	gibberellic acid-insensitive protein
GAPC-2	glyceraldehyde 3-phosphate dehydrogenase C-2
GO	gene ontology
GRAS	GA INSENSITIVE (GAI), REPRESSOR of ga1-3 (RGA), and SCARECROW (SCR) protein domain
GTP	guanosine-5′-triphosphate
HvRBK	Hordeum vulgare ROP-binding kinase
LAX	Like auxin1
MtRRK1	*Medicago truncatula* ROP-activated Receptor-like Kinase 1
NPF	NRT1/PTR family
NRT	nitrate transporter
PCR	polymerase chain reaction
PIF	phytochrome interacting factor
PIN	PINNOID protein
PTR	peptide transporter
RLCK	receptor-like cytoplasmic kinase
RLK	receptor-like kinase
ROP	Rho-of-plants
SD	short day
T-DNA	transfer-DNA
TF	transcription factor

References

1. Lehti-Shiu, M.D.; Shiu, S.-H. Diversity, classification and function of the plant protein kinase superfamily. *Philos. Trans. R. Soc. B Biol. Sci.* **2012**, *367*, 2619–2639. [CrossRef] [PubMed]
2. Shiu, S.H.; Bleecker, A.B. Receptor-like kinases from Arabidopsis form a monophyletic gene family related to animal receptor kinases. *Proc. Natl. Acad. Sci. USA* **2001**, *98*, 10763–10768. [CrossRef] [PubMed]
3. Shiu, S.-H.; Karlowski, W.M.; Pan, R.; Tzeng, Y.-H.; Mayer, K.F.X.; Li, W.-H. Comparative analysis of the receptor-like kinase family in Arabidopsis and rice. *Plant Cell* **2004**, *16*, 1220–1234. [CrossRef] [PubMed]
4. Liang, X.; Zhou, J.-M. Receptor-Like Cytoplasmic Kinases: Central players in plant receptor kinase–mediated signaling. *Annu. Rev. Plant Biol.* **2018**, *69*, 267–299. [CrossRef] [PubMed]
5. Jurca, M.E.; Bottka, S.; Fehér, A. Characterization of a family of Arabidopsis receptor-like cytoplasmic kinases (RLCK class VI). *Plant Cell Rep.* **2008**, *27*, 739–748. [CrossRef]
6. Huesmann, C.; Reiner, T.; Hoefle, C.; Preuss, J.; Jurca, M.E.; Domoki, M.; Fehér, A.; Hückelhoven, R. Barley ROP binding kinase1 is involved in microtubule organization and in basal penetration resistance to the barley powdery mildew fungus. *Plant Physiol.* **2012**, *159*, 311–320. [CrossRef]

7. Molendijk, A.J.; Ruperti, B.; Singh, M.K.; Dovzhenko, A.; Ditengou, F.A.; Milia, M.; Westphal, L.; Rosahl, S.; Soellick, T.; Uhrig, J.; et al. A cysteine-rich receptor-like kinase NCRK and a pathogen-induced protein kinase RBK1 are Rop GTPase interactors. *Plant J.* **2008**, *53*, 909–923. [CrossRef]

8. Dorjgotov, D.; Jurca, M.E.; Fodor-Dunai, C.; Szűcs, A.; Ötvös, K.; Klement, É.; Bíró, J.; Fehér, A. Plant Rho-type (Rop) GTPase-dependent activation of receptor-like cytoplasmic kinases in vitro. *FEBS Lett.* **2009**, *583*, 1175–1182. [CrossRef]

9. Reiner, T.; Hoefle, C.; Huesmann, C.; Ménesi, D.; Fehér, A.; Hückelhoven, R. The Arabidopsis ROP-activated receptor-like cytoplasmic kinase RLCK VI_A3 is involved in control of basal resistance to powdery mildew and trichome branching. *Plant Cell Rep.* **2015**, *34*, 457–468. [CrossRef]

10. Zhao, Z.; Manser, E. PAK and other Rho-associated kinases–effectors with surprisingly diverse mechanisms of regulation. *Biochem. J.* **2005**, *386*, 201–214. [CrossRef]

11. Fehér, A.; Lajkó, D.B. Signals fly when kinases meet Rho-of-plants (ROP) small G-proteins. *Plant Sci.* **2015**, *237*, 93–107. [CrossRef] [PubMed]

12. Lajkó, D.B.; Valkai, I.; Domoki, M.; Ménesi, D.; Ferenc, G.; Ayaydin, F.; Fehér, A. In silico identification and experimental validation of amino acid motifs required for the Rho-of-plants GTPase-mediated activation of receptor-like cytoplasmic kinases. *Plant Cell Rep.* **2018**, *37*, 627–639. [CrossRef]

13. Feiguelman, G.; Fu, Y.; Yalovsky, S. ROP GTPases structure-function and signaling pathways. *Plant Physiol.* **2018**, *176*, 57–79. [CrossRef] [PubMed]

14. Enders, T.A.; Frick, E.M.; Strader, L.C. An Arabidopsis kinase cascade influences auxin-responsive cell expansion. *Plant J.* **2017**, *92*, 68–81. [CrossRef]

15. Kleinboelting, N.; Huep, G.; Kloetgen, A.; Viehoever, P.; Weisshaar, B. GABI-Kat SimpleSearch: New features of the Arabidopsis thaliana T-DNA mutant database. *Nucleic Acids Res.* **2012**, *40*, D1211–D1215. [CrossRef] [PubMed]

16. Scholl, R.L.; May, S.T.; Ware, D.H. Seed and molecular resources for Arabidopsis. *Plant Physiol.* **2000**, *124*, 1477–1480. [CrossRef] [PubMed]

17. Rubio, V.; Shen, Y.; Saijo, Y.; Liu, Y.; Gusmaroli, G.; Dinesh-Kumar, S.P.; Deng, X.W. An alternative tandem affinity purification strategy applied to Arabidopsis protein complex isolation. *Plant J.* **2005**, *41*, 767–778. [CrossRef]

18. Zuo, J.; Niu, Q.W.; Chua, N.H. Technical advance: An estrogen receptor-based transactivator XVE mediates highly inducible gene expression in transgenic plants. *Plant J.* **2000**, *24*, 265–273. [CrossRef] [PubMed]

19. Ivakov, A.; Persson, S. Plant cell shape: Modulators and measurements. *Front. Plant Sci.* **2013**, *4*, 439. [CrossRef]

20. Yamaguchi, S. Gibberellin metabolism and its regulation. *Annu. Rev. Plant Biol.* **2008**, *59*, 225–251. [CrossRef]

21. Choi, H.; Jeong, S.; Kim, D.S.; Na, H.J.; Ryu, J.S.; Lee, S.S.; Nam, H.G.; Lim, P.O.; Woo, H.R. The homeodomain-leucine zipper ATHB23, a phytochrome B-interacting protein, is important for phytochrome B-mediated red light signaling. *Physiol. Plant.* **2014**, *150*, 308–320. [CrossRef]

22. Hedden, P.; Thomas, S.G. Gibberellin biosynthesis and its regulation. *Biochem. J.* **2012**, *444*, 11–25. [CrossRef]

23. Wulff, N.; Ernst, H.A.; Jørgensen, M.E.; Lambertz, S.; Maierhofer, T.; Belew, Z.M.; Crocoll, C.; Motawia, M.S.; Geiger, D.; Jørgensen, F.S.; et al. An optimized screen reduces the number of GA transporters and provides insights into Nitrate Transporter 1/Peptide Transporter family substrate determinants. *Front. Plant Sci.* **2019**, *10*, 1106. [CrossRef] [PubMed]

24. Chiba, Y.; Shimizu, T.; Miyakawa, S.; Kanno, Y.; Koshiba, T.; Kamiya, Y.; Seo, M. Identification of Arabidopsis thaliana NRT1/PTR FAMILY (NPF) proteins capable of transporting plant hormones. *J. Plant Res.* **2015**, *128*, 679–686. [CrossRef] [PubMed]

25. Tal, I.; Zhang, Y.; Jørgensen, M.E.; Pisanty, O.; Barbosa, I.C.R.; Zourelidou, M.; Regnault, T.; Crocoll, C.; Erik Olsen, C.; Weinstain, R.; et al. The Arabidopsis NPF3 protein is a GA transporter. *Nat. Commun.* **2016**, *7*, 11486. [CrossRef] [PubMed]

26. Corratgé-Faillie, C.; Lacombe, B. Substrate (un)specificity of Arabidopsis NRT1/PTR FAMILY (NPF) proteins. *J. Exp. Bot.* **2017**, *68*, 3107–3113. [CrossRef]

27. Kanno, Y.; Oikawa, T.; Chiba, Y.; Ishimaru, Y.; Shimizu, T.; Sano, N.; Koshiba, T.; Kamiya, Y.; Ueda, M.; Seo, M. AtSWEET13 and AtSWEET14 regulate gibberellin-mediated physiological processes. *Nat. Commun.* **2016**, *7*, 13245. [CrossRef]

28. Sun, T.P.; Kamiya, Y. The Arabidopsis GA1 locus encodes the cyclase ent-kaurene synthetase A of gibberellin biosynthesis. *Plant Cell* **1994**, *6*, 1509–1518.

29. Archacki, R.; Buszewicz, D.; Sarnowski, T.J.; Sarnowska, E.; Rolicka, A.T.; Tohge, T.; Fernie, A.R.; Jikumaru, Y.; Kotlinski, M.; Iwanicka-Nowicka, R.; et al. BRAHMA ATPase of the SWI/SNF chromatin remodeling complex acts as a positive regulator of gibberellin-mediated responses in arabidopsis. *PLoS ONE* **2013**, *8*, e58588. [CrossRef]

30. Li, K.; Yu, R.; Fan, L.-M.; Wei, N.; Chen, H.; Deng, X.W. DELLA-mediated PIF degradation contributes to coordination of light and gibberellin signalling in Arabidopsis. *Nat. Commun.* **2016**, *7*, 11868. [CrossRef]

31. Oh, E.; Zhu, J.Y.; Bai, M.Y.; Arenhart, R.A.; Sun, Y.; Wang, Z.Y. Cell elongation is regulated through a central circuit of interacting transcription factors in the Arabidopsis hypocotyl. *eLife* **2014**, *2014*, e03031. [CrossRef] [PubMed]

32. Vij, S.; Giri, J.; Dansana, P.K.; Kapoor, S.; Tyagi, A.K. The receptor-like cytoplasmic kinase (OsRLCK) gene family in rice: Organization, phylogenetic relationship, and expression during development and stress. *Mol. Plant* **2008**, *1*, 732–750. [CrossRef]

33. Lin, W.; Ma, X.; Shan, L.; He, P. Big Roles of Small Kinases: The Complex Functions of Receptor-like Cytoplasmic Kinases in Plant Immunity and Development. *J. Integr. Plant Biol.* **2013**, *55*, 1188–1197. [CrossRef] [PubMed]

34. Chen, X.; Wu, S.; Liu, Z.; Friml, J. Environmental and endogenous control of cortical microtubule orientation. *Trends Cell Biol.* **2016**, *26*, 409–419. [CrossRef] [PubMed]

35. Ma, Q.; Wang, X.; Sun, J.; Mao, T. Coordinated regulation of hypocotyl cell elongation by light and ethylene through a microtubule destabilizing protein. *Plant Physiol.* **2018**, *176*, 678–690. [CrossRef]

36. Fu, Y.; Xu, T.; Zhu, L.; Wen, M.; Yang, Z. A ROP GTPase signaling pathway controls cortical microtubule ordering and cell expansion in Arabidopsis. *Curr. Biol.* **2009**, *19*, 1827–1832. [CrossRef]

37. Cosgrove, D.J. Loosening of plant cell walls by expansins. *Nature* **2000**, *407*, 321–326. [CrossRef]

38. Van Sandt, V.S.T.; Suslov, D.; Verbelen, J.-P.; Vissenberg, K. Xyloglucan endotransglucosylase activity loosens a plant cell wall. *Ann. Bot.* **2007**, *100*, 1467–1473. [CrossRef]

39. Sadava, D.; Chrispeels, M.J. Hydroxyproline-rich cell wall protein (extensin): Role in the cessation of elongation in excised pea epicotyls. *Dev. Biol.* **1973**, *30*, 49–55. [CrossRef]

40. Passardi, F.; Penel, C.; Dunand, C. Performing the paradoxical: How plant peroxidases modify the cell wall. *Trends Plant Sci.* **2004**, *9*, 534–540. [CrossRef]

41. Bernardo-García, S.; de Lucas, M.; Martínez, C.; Espinosa-Ruiz, A.; Davière, J.-M.; Prat, S. BR-dependent phosphorylation modulates PIF4 transcriptional activity and shapes diurnal hypocotyl growth. *Genes Dev.* **2014**, *28*, 1681–1694. [CrossRef] [PubMed]

42. Willige, B.C.; Isono, E.; Richter, R.; Zourelidou, M.; Schwechheimer, C. Gibberellin regulates PIN-FORMED abundance and is required for auxin transport–dependent growth and development in Arabidopsis thaliana. *Plant Cell* **2011**, *23*, 2184–2195. [CrossRef]

43. Curtis, M.D.; Grossniklaus, U. A gateway cloning vector set for high-throughput functional analysis of genes in planta. *Plant Physiol.* **2003**, *133*, 462–469. [CrossRef] [PubMed]

44. Koncz, C.; Martini, N.; Szabados, L.; Hrouda, M.; Bachmair, A.; Schell, J. Specialized vectors for gene tagging and expression studies. In *Plant Molecular Biology Manual*; Gelvin, S.B., Schilperoort, R.A., Eds.; Springer: Dordrecht, The Netherland, 1994; pp. 53–74, ISBN 978-94-011-0511-8.

45. Clough, S.J.; Bent, A.F. Floral dip: A simplified method for Agrobacterium -mediated transformation of Arabidopsis thaliana. *Plant J.* **1998**, *16*, 735–743. [CrossRef]

46. Schneider, C.A.; Rasband, W.S.; Eliceiri, K.W. NIH Image to ImageJ: 25 years of image analysis. *Nat. Methods* **2012**, *9*, 671–675. [CrossRef] [PubMed]

47. Talbot, M.J.; White, R.G. Cell surface and cell outline imaging in plant tissues using the backscattered electron detector in a variable pressure scanning electron microscope. *Plant Methods* **2013**, *9*, 40. [CrossRef]

48. Urbanová, T.; Tarkowská, D.; Novák, O.; Hedden, P.; Strnad, M. Analysis of gibberellins as free acids by ultra performance liquid chromatography-tandem mass spectrometry. *Talanta* **2013**, *112*, 85–94. [CrossRef]

49. Rittenberg, D.; Foster, G.L. A new procedure for quantitative analysis by isotope dilution, with application to the determination of amino acids and fatty acids. *J. Biol. Chem.* **1940**, *133*, 737–744.

50. Wu, G.; Carville, J.S.; Spalding, E.P. ABCB19-mediated polar auxin transport modulates Arabidopsis hypocotyl elongation and the endoreplication variant of the cell cycle. *Plant J.* **2016**, *85*, 209–218. [CrossRef]

51. Ortuño, A.; Río, J.D.D.; Casas, J.; Serrano, M.; Acosta, M.; Sánchez-Bravo, J. Influence of ACC and Ethephon on cell growth in etiolated lupin hypocotyls. dependence on cell growth state. *Biol. Plant.* **2008**, *33*, 81. [CrossRef]

52. Alabadí, D.; Gil, J.; Blázquez, M.A.; García-Martínez, J.L. Gibberellins repress photomorphogenesis in darkness. *Plant Physiol.* **2004**, *134*, 1050–1057. [CrossRef] [PubMed]

53. Baba, A.I.; Valkai, I.; Labhane, N.M.; Koczka, L.; Andrási, N.; Klement, É.; Darula, Z.; Medzihradszky, K.F.; Szabados, L.; Fehér, A.; et al. CRK5 Protein Kinase Contributes to the Progression of Embryogenesis of Arabidopsis thaliana. *Int. J. Mol. Sci.* **2019**, *20*, 6120. [CrossRef] [PubMed]

54. Bolger, A.M.; Lohse, M.; Usadel, B. Trimmomatic: A flexible trimmer for Illumina sequence data. *Bioinformatics* **2014**, *30*, 2114–2120. [CrossRef]

55. Kim, D.; Pertea, G.; Trapnell, C.; Pimentel, H.; Kelley, R.; Salzberg, S.L. TopHat2: Accurate alignment of transcriptomes in the presence of insertions, deletions and gene fusions. *Genome Biol.* **2013**, *14*, R36. [CrossRef] [PubMed]

56. Cheng, C.-Y.; Krishnakumar, V.; Chan, A.P.; Thibaud-Nissen, F.; Schobel, S.; Town, C.D. Araport11: A complete reannotation of the Arabidopsis thaliana reference genome. *Plant J.* **2017**, *89*, 789–804. [CrossRef] [PubMed]

MDPI
St. Alban-Anlage 66
4052 Basel
Switzerland
Tel. +41 61 683 77 34
Fax +41 61 302 89 18
www.mdpi.com

IJMS Editorial Office
E-mail: ijms@mdpi.com
www.mdpi.com/journal/ijms